Urartu Özgür Şafak Şeker (Ed.)
Synthetic Biology for Therapeutics

Also of interest

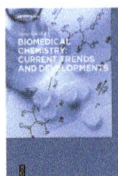

Biomedical Chemistry:
Current Trends and Developments
Nuno Vale, 2015
ISBN 978-3-11-046874-8, e-ISBN 978-3-11-046875-5

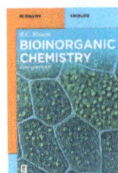

Bioinorganic Chemistry.
Some New Facets
Ram Charitra Maurya, 2021
ISBN 978-3-11-072729-6, e-ISBN 978-3-11-072730-2

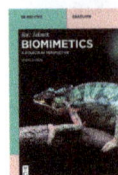

Biomimetics.
A Molecular Perspective
Raz Jelinek, 2021
ISBN 978-3-11-070944-5, e-ISBN 978-3-11-070949-0

Pharmaceutical Chemistry.
Drug Design and Action
Joaquín M. Campos Rosa, 2024
ISBN 978-3-11-131654-3, e-ISBN 978-3-11-131690-1

Synthetic Biology for Therapeutics

Engineering Cells for Living Drugs

Edited by
Urartu Özgür Şafak Şeker

DE GRUYTER

Editor
Urartu Özgür Şafak Şeker, Ph.D.
National Nanotechnology Research Center (UNAM)
Bilkent University
Ankara 06800
Turkey
urartu@bilkent.edu.tr

ISBN 978-3-11-132948-2
e-ISBN (PDF) 978-3-11-132949-9
e-ISBN (EPUB) 978-3-11-132962-8

Library of Congress Control Number: 2024943047

Bibliographic information published by the Deutsche Nationalbibliothek
The Deutsche Nationalbibliothek lists this publication in the Deutsche Nationalbibliografie;
detailed bibliographic data are available on the internet at http://dnb.dnb.de.

© 2025 Walter de Gruyter GmbH, Berlin/Boston
Cover image: Design Cells/iStock/Getty Images Plus
Typesetting: Integra Software Services Pvt. Ltd.

www.degruyter.com

——

This book is dedicated to the memory of Mustafa Kemal Ataturk and the first centenary of the Turkish Republic founded by him, celebrating its rich history.

Dr. Şeker dedicates this work to his mother Guldane Şeker and his father Casim Şeker.

Contents

List of contributing authors —— XI

Recep Erdem Ahan, Derin Akman, Ece Avcı,
and Urartu Özgür Şafak Şeker
Chapter 1
Programming cells with synthetic biology —— 1

Merve Yavuz, Ahmet Hınçer, Aslı Semerci,
and Urartu Özgür Şafak Şeker
Chapter 2
Biological devices for cellular targeting and decision-making —— 31

Melis Karaca, Senem Şen, Fayiti Fuerkaiti,
and Urartu Özgür Şafak Şeker
Chapter 3
Engineering mammalian cell for cancer —— 59

Doğuş Akboğa, Aslı Doğruer, Damla Albayrak, Gozeel Binte Shahid,
and Urartu Özgür Şafak Şeker
Chapter 4
Engineering microbial cells for cancer —— 87

Nazlıcan Tunç, Mehmet Emin Bakar, Burak Çalışkan, Tolga Tarkan Ölmez,
and Urartu Özgür Şafak Şeker
Chapter 5
New generation cellular engineering for living therapeutics —— 119

Index —— 171

List of contributing authors

Recep Erdem Ahan
UNAM–Institute of Materials Science and
Nanotechnology
Bilkent University
Ankara 06800
Turkey

Ece Avcı
UNAM–Institute of Materials Science and
Nanotechnology
Bilkent University
Ankara 06800
Turkey

Derin Akman
UNAM–Institute of Materials Science and
Nanotechnology
Bilkent University
Ankara 06800
Turkey

Merve Yavuz
UNAM–Institute of Materials Science and
Nanotechnology
Bilkent University
Ankara 06800
Turkey

Ahmet Hınçer
UNAM–Institute of Materials Science and
Nanotechnology
Bilkent University
Ankara 06800
Turkey

Aslı Semerci
UNAM–Institute of Materials Science and
Nanotechnology
Bilkent University
Ankara 06800
Turkey

Melis Karaca
UNAM–Institute of Materials Science and
Nanotechnology
Bilkent University
Ankara 06800
Turkey

Senem Şen
UNAM–Institute of Materials Science and
Nanotechnology
Bilkent University
Ankara 06800
Turkey

Fayit Fuerkati
UNAM–Institute of Materials Science and
Nanotechnology
Bilkent University
Ankara 06800
Turkey

Doğuş Akboğa
UNAM–Institute of Materials Science and
Nanotechnology
Bilkent University
Ankara 06800
Turkey

Aslı Doğruer
UNAM–Institute of Materials Science and
Nanotechnology
Bilkent University
Ankara 06800
Turkey

Damla Albayrak
UNAM–Institute of Materials Science and
Nanotechnology
Bilkent University
Ankara 06800
Turkey

https://doi.org/10.1515/9783111329499-203

Gozeel Binte Shadid
UNAM–Institute of Materials Science and
Nanotechnology
Bilkent University
Ankara 06800
Turkey

Urartu Özgür Şafak Şeker
UNAM–Institute of Materials Science and
Nanotechnology
Bilkent University
Ankara 06800
Turkey

Nazlıcan Tunç
UNAM–Institute of Materials Science and
Nanotechnology
Bilkent University
Ankara 06800
Turkey

Mehmet Emin Bakar
UNAM–Institute of Materials Science and
Nanotechnology
Bilkent University
Ankara 06800
Turkey

Burak Çalışkan
Department of Biomedical Engineering
TOBB University of Economics and Technology
Ankara 06510
Turkey

Tolga Tarkan Ölmez
Department of Biomedical Engineering
Başkent University
Ankara 06790
Turkey

Recep Erdem Ahan, Derin Akman, Ece Avcı,
and Urartu Özgür Şafak Şeker

Chapter 1
Programming cells with synthetic biology

Abstract: Biological systems are the most advanced molecular machines known to humans. Cells can execute intricate actions via proteins encoded in their genomes. Their immense capabilities are rooted in the billions of years of evolution wherein organisms and biological molecules have been diversified by natural selection to adapt to continuously changing ecological conditions in order to ensure survival. Evolution has yielded many useful functions of biological systems that can be exploited for human use. However, wild-type cells and biological molecules are suboptimal for specific applications because their capabilities are shaped and dictated for survival, not for the requirements of any human application. Synthetic biology approaches offer to augment the inherent competence of cells through genetic reprogramming, wherein cells are considered a collection of biomolecular modules composed of biological parts. Based on this approach, novel molecular abilities can be constructed by reusing biological parts to build synthetic genetic modules implemented in cellular hosts. In this chapter, reprogramming cells by building synthetic genetic modules will be summarized. In the first section, molecular methodologies, including de novo DNA synthesis, cloning, and genome engineering, will be briefly mentioned. Then, genetic modules for signal sensing and signal processing will be explained.

1.1 Introduction

Living cells are astonishing microrobots that can perform many molecular functions simultaneously. In addition to their self-replicative abilities, cells can sense their environment through their receptors, compute the sensed signals, and produce mechanical and/or chemical outputs based on the signals. Moreover, in a multicellular organism, the whole-body system, including organs and circulating cells that have completely irrelevant molecular functions, can be differentiated from a single-celled zygote. These mind-blowing capabilities and complexities of living cells emerge from the molecular interactions of biomolecules within the cellular membrane. In essence, cells are composed of a bunch of molecules that are trapped in a lipid bilayer membrane. However, the interactions of these molecules are specific, which leads to the emergence of meaningful biochemical functions.

Recep Erdem Ahan, Derin Akman, Ece Avcı, and Urartu Özgür Şafak Şeker, UNAM–Institute of
Materials Science and Nanotechnology, Bilkent University

https://doi.org/10.1515/9783111329499-001

The information that dictates the functions of cells is encrypted in their genome. During the information flow and decryption of genomic information inside the cells, the genome is transcribed from DNA molecules to RNA molecules, and proteins are produced based on the sequence of the RNA that is transcribed. The biological functions are mainly exerted by protein molecules, which function as catalysts for biochemical processes. Aside from catalysis, biological processes are regulated temporally and spatially with other proteins as well as RNA and small molecules through direct and/or indirect interactions [1]. These interaction maps of molecules within and outside of the cells form networks in which cells decide how to react (produce an output).

During the course of evolution, cells have exchanged genetic information through horizontal transfer; moreover, DNA sequences found within cells constantly mutate because of environmental factors such as radiation [2]. These continuous changes in genomes have given rise to the formation of new proteins and RNA molecules, which have inevitably resulted in the origination of novel networks and biological functions. Over approximately 3 billion years of biological evolution, life on the Earth has been adapted to survive distinctly diverse ecological niches such as animal guts [3], salty lakes [4], and volcanic caves [5]. In order to occupy even the extreme conditions, cells have evolved to acquire unique molecular machineries. For instance, thermophilic organisms possess distinct cellular membranes that are enriched with ether-linked lipids and thermostable proteins with higher amounts of hydrophobic amino acid compositions [6, 7]. On the other hand, psychrophilic organisms encode proteins with polar and noncharged amino acids to cope with cold conditions [8]. By selecting the fittest, natural selection has led to the evolution of biomolecules with different biochemical functions as well as biomolecules with the same functions but distinct working conditions such as temperature and salt concentration. As a result, the Earth contains immense amounts of biodiversity.

This biodiversity on the Earth can be exploited for human use. For instance, microorganisms can be used to degrade certain chemicals in the environment [9, 10], or to regulate the immune system in the body [11], or to produce valuable chemicals [12]. However, most of the cellular systems and biomolecules found in nature are suboptimal in their native form for the applications in which they can be utilized [13]. The main reason for this is that cells found in nature haven't evolved to function in a particular application but rather have evolved to survive a particular environment. In addition, for certain applications, the required cellular functions have not evolved in any single cell type. Instead, the necessary biological molecules are encoded across different species [14, 15]. For these reasons, wild-type cells are limited to the capabilities that they have acquired during evolution.

Reprogramming and engineering cells with recombinant technologies open a new avenue for the use of biological systems for technological applications. Although there are early examples of biotechnological applications, such as the recombinant production of insulin [16] after the discovery of PCR and DNA ligation techniques, the genetic programming of cells bloomed after the advancements in technologies such as

DNA construction and sequencing and the adaptation of synthetic biology approaches in the mid-2000s [17–19].

In synthetic biology, cells are simplified as modules (also referred to as genetic circuits) composed of biological parts (also referred to as genetic parts). Biological parts are the smallest functional blocks within cells. Transcription factor (TF) binding sequences, regulatory RNA sequences, protein-coding sequences, peptide tag sequences, etc., can be given as examples of biological parts [19]. Meanwhile, modules are logical combinations of biological parts to generate biological operations in the cells such as signal sensing, signal processing, and response generation. To genetically reprogram cells with synthetic biology, these modules are redesigned and constructed with modular biological parts that can provide desired biological operations [20]. The constructed genetic modules are tested within a base cellular host that provides side biological processes for desired functions. Owing to the modularity of biological parts that are used, the genetic modules can be redesigned with biological parts that have the same function if the initial module design fails. This "design-build-test-learn" engineering approach enables logical design and construction of optimal cellular systems for an application [21]. To better understand concepts in synthetic biology, an airplane Lego analogy can be useful. In this analogy, biological parts, modules, wild-type cells, and synthetically designed cells can be considered as individual Lego parts, parts of the airplane such as wings, a misshaped airplane, and realistically imitated Lego airplane, respectively.

Herein this chapter, we will summarize the current DNA technologies to construct synthetic genetic modules. In addition, native and synthetic sensing and signal processing modules that are highly utilized in synthetic biology will be summarized.

1.2 DNA technologies for cell reprogramming

The synthesis and construction of synthetic genetic modules heavily relies on de novo synthesis of DNA encoding biological parts in vitro [22]. The synthesized biological parts are assembled via molecular cloning techniques into a shuttle vector plasmid. The assembled vector DNA is then transferred to the cellular host, which can be either a prokaryotic cell, such as a strain of *Escherichia coli* bacterium, or a eukaryote cell, such as a mammalian cell line, or a strain of *Saccharomyces cerevisiae*. To transfer vector plasmids inside cells, different methodologies have been optimized for each species including the use of physical stress, viral particles, and lipid nanoparticles. In some cases, the functionality of genetic modules is hindered due to the inherent gene networks within the base cellular host. Genome modification tools such as the CRISPR-Cas system have been deployed to resolve such issues. In this section, methodologies that are utilized in recombinant DNA technology will be summarized.

1.2.1 De novo DNA synthesis

Current de novo DNA synthesis strategies are dependent on the enzymatic assembly of short oligonucleotides obtained via chemical synthesis. For the chemical synthesis, the seeding nucleotide is immobilized on a solid support, and the phosphoramidite derivative of individual nucleotides is added one by one to each other in a sequential manner to get the oligonucleotide with the desired DNA sequence [23]. To further decrease the cost and time, silicon chips are constructed with individual microscopic reaction sites wherein multiple oligonucleotide synthesis can be done spontaneously [24, 25]. However, phosphoramidite-based oligonucleotide synthesis (POS) can be used effectively to synthesize oligonucleotides smaller than 200 bp due to the reaction yield being limited to the inherent chemistry. Therefore, enzymatic assembly of short oligonucleotides is necessary to get coding DNA sequences for biological parts, which are typically larger than 200 bp [23].

For the assembly of short oligonucleotides obtained from the chemical synthesis, two main methodologies are utilized, namely polymerase cycling assembly (PCA) and isothermal DNA assembly (iDA) (popularly also known as Gibson assembly). In the PCA method [26], the synthesized gene is divided into short, overlapping oligonucleotides that alternate between both sense and antisense strands of the gene. Following the mixing and hybridization of oligonucleotides, one duplex DNA fragment with gaps between oligonucleotides is formed. High-fidelity polymerase, such as Pfu polymerase, is used to fill the gaps and synthesize DNA encoding the full gene sequence. Primers that attach to 5′ ends of each strand are then used to amplify the final desired DNA sequence. Although the PCA method is hindered by the presence of repeating sequences, typically, it is possible to obtain 2.5 kb long DNA fragments from short chemically synthesized oligonucleotides (Figure 1.1).

The iDA can be used to synthesize large genes by snitching smaller gene fragments together. The iDA reaction contains three enzymes, which are high-fidelity DNA polymerase, T5′ exonuclease, and Taq ligase, along with the free nucleotides and cofactors [27]. Multiple gene fragments with complementary overlapping regions at their terminal can be stitched together with an iDA reaction. The 5′ ends of each strand of each fragment gene are degraded by the T5′ exonuclease enzyme, generating sticky 3′ ends. The sticky 3′ ends hybridize during the reaction, and gaps are filled by a high-fidelity enzyme. Lastly, the remaining nicks are repaired by the Taq ligase enzyme. To synthesize large genes (>2.5 kb), smaller fragment genes obtained from the PCA method can be assembled with iDA; however, iDA can also be utilized to synthesize genes from chemically synthesized oligonucleotides. In contrast to the PCA method, oligonucleotides should be hybridized perfectly without any gap in the assembly with iDA. The nicks in the hybridized duplex DNA are sealed with Taq ligase enzyme. In addition to amplifying the desired sequence with PCR, the assembled gene product can be cloned directly into a cloning vector with iDA (Figure 1.1).

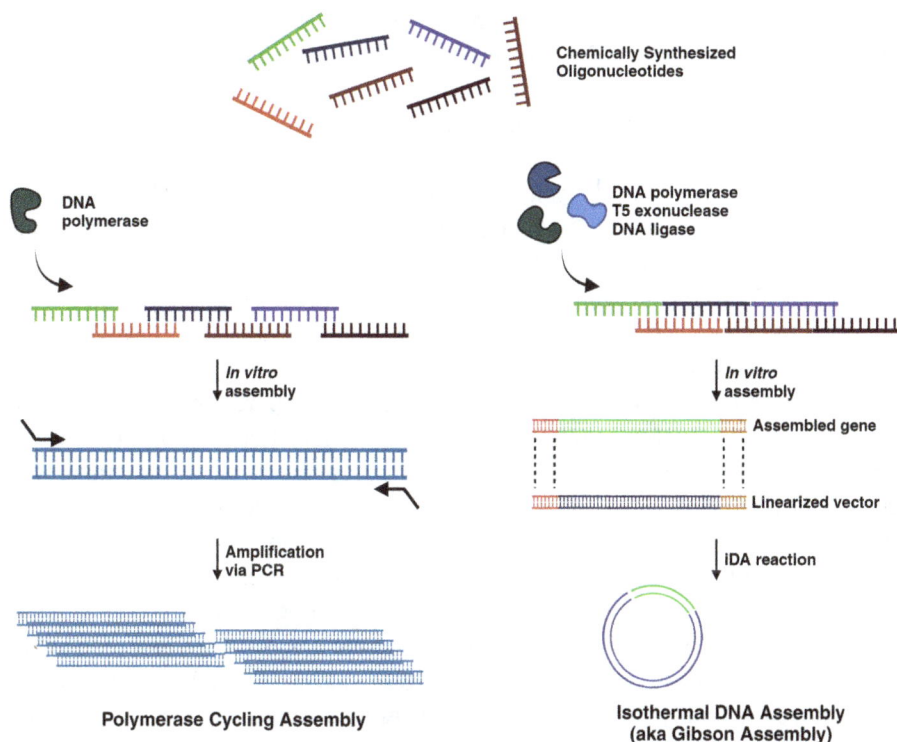

Figure 1.1: Depiction of PCA (left) and the iDA (right) for de novo gene synthesis from chemically synthesized oligonucleotides (created with BioRender.com).

The DNA synthesis services are centralized and generally provided by vendors such as GenScript, Integrated DNA Technologies, and Twist Biosciences due to the labor-intensive workflow and initial establishment cost for the facility required for the POS. However, the emerging technological developments in DNA synthesis, such as enzymatic de novo DNA synthesis and automation, can further decrease the costs and skills required, which can provide democratization of DNA synthesis from centralized vendors, thereby accelerating synthetic biology research.

1.2.2 Molecular cloning and recombinant vector plasmids

DNA molecules encoding synthetic genetic modules are inserted into recombinant vector plasmids via molecular cloning techniques to transfer the DNA molecules into cellular hosts. Plasmids are extrachromosomal DNA molecules found generally in bacteria that have the necessary elements for autonomous replication and propagation. For recombinant technologies, natural plasmids are repurposed and designed as vectors for delivering foreign DNA into host cells [28–31]. Plasmid vectors share several common features to ease the burden on the selection of cells carrying the plasmid and molecular construction of the plasmid with desired sequences. For instance, plasmid vectors contain an ori region for propagation into the recipient cell host and an antibiotic resistance gene to select cells carrying the plasmid vector DNA. In addition to these elements, promoters for protein expression, multiple cloning sites with different restriction enzymes, OriT region for conjugation, and other elements can be incorporated into plasmid vectors DNA based on the applications in which the plasmid vectors are going to be used [32, 33]. Because of the necessity to change each component of plasmid vectors in synthetic biology applications, many modular plasmid vectors have been designed such as pZ vector series [34], BioBricks [35], pSeva [36], and pMTL80000 series [37].

To assemble plasmid vectors with desired biological parts, several techniques have been developed. For instance, the iDA reaction [27] described in the previous section can be utilized to assemble vector plasmids from the sketch by combining synthetic DNA fragments. However, the DNA fragments with high sequence similarities or repetitive sequences at the ends of strands cannot be combined with the iDA reaction because of the homology-based nature of the iDA. To overcome the obstacles of iDA reaction, Golden Gate Cloning techniques can be used. In Golden Gate Cloning [38], type IIS restriction enzymes are utilized to create unique, short, and compatible sticky ends. These ends can be ligated with ligase enzymes. Type IIS restriction enzymes have the ability to cleave DNA sequences outside of their recognition sites; therefore, the recognition sites can be removed after the restriction reaction.

Designing DNA fragments for both the iDA reaction and Golden Gate Cloning reaction is vital for successful assembly. Therefore, in silico plasmid design and visualization tools, such as Benchling, Geneious, and SnapGene, have been developed. These software packages provide built-in tools that can be used to visualize and annotate plasmid sequences. In addition, experimental design to assemble a plasmid from DNA fragments can be performed in silico for the iDA reaction and Golden Gate Cloning wherein required PCR reactions and primers are generated by the software.

1.2.3 Recombinant DNA transfer into cell hosts

Genetic reprogramming of cells necessitates the transfer of foreign DNA, which encodes synthetic biological modules, into the cytosol. Since the DNA and cellular membrane are negatively charged, different methods are employed to induce the foreign DNA crossing through the cellular membrane. These methods can be categorized into three categories: physical, chemical, and biological [39].

Physical DNA transfer methods involve the induction of temporary pores on the cellular membrane via force wherein DNA can diffuse into the cytosol. In general, physical transfer methods do not utilize any chemical and/or biological agents; therefore, they are safer and user-friendly. On the other hand, dedicated laboratory equipment generates the physical force needed for the protocols. As an example, electroporation [40] is a popular method to transfer DNA into different types of cells, including bacteria and mammalian cells [41]. In this method, a strong electric field is applied on the cells, thereby causing the formation of transient holes on the cellular membrane. The transient holes allow the diffusion of recombinant DNA into the cytosol. Although transient holes induced by electroporation do not harm cells in the optimized conditions, the electroporation parameters such as pulse length, electric field strength, and duration should be optimized for each type of cell, which is a cumbersome process [42]. Aside from electroporation, microneedle injection [43], gene gun [44], sonoporation [45], laser-based transfection [46], and magnetofection [47] are the methodologies that utilize physical forces to transfect cells with recombinant DNA.

In chemical DNA transfer methods, negatively charged DNA molecules are complexed with positively charged carriers such as Ca^{2+} ions [48], polymers [49], and lipids [49]. The DNA-carrier complexes can bind and pass through the negatively charged cellular membrane. Compared to physical gene transfer methods, chemical methods are cheaper because these methods do not require any specialized instruments [39]. In addition, chemical gene transfer methods can be used to transfer DNA to cells not only in vitro but also in vivo. For instance, lipid mixtures are utilized to create liposomes, which are spherical vesicles on a nanometric scale with lipid bilayers. DNA molecules are encapsulated within the liposomes, thereby protecting them from environmental stresses and biological degradation. In addition, many of the lipids used in liposome production are shown to be not toxic for mammalian cells [50].

Biological DNA transfer methods, including viral transduction and bacterial conjugation, exploit the natural ability of biological systems. In the viral transduction method, viruses are genetically designed to deliver recombinant DNA into the cellular host. As viruses utilize specific receptors to invade their cognate hosts, many viruses exhibit narrow tropism, targeting only certain types of cells for the delivery of recombinant DNA [51]. Additionally, it is hypothesized that a virus exists for every living organism on the Earth due to the co-evolution of hosts and viruses and the numerous biodiversity of viruses [52]. Consequently, a biological DNA shuttle can be engineered for each individual living organism. On the other hand, the bacterial conjugation method exploits the horizontal gene transfer mechanism between bacteria species [53]. Conjugative plasmids are redesigned to carry the recombinant DNA fragment and utilized to transfer the DNA from model donor species such as *E. coli* to the recipient cellular host [37]. Bacterial conjugation is generally preferred when the chemical and physical DNA transfer method is not feasible for the recipient cells, such as anaerobic bacteria species. Similar to liposomes, biological DNA transfer methods can be used to deliver recombinant DNA into cells in vivo.

In summary, DNA transfer methods can be categorized into three groups, which are physical, chemical, and biological methods. Each methodology has its own advantages and disadvantages. The appropriate methods should be selected based on the specific cells and applications that cells are being designed for.

1.2.4 Genome engineering with CRISPR-Cas systems

Synthetic genetic modules are implemented into an appropriate cellular host for functioning and provide the necessary side processes such as transcription, translation, and posttranslational modifications. In addition, the inherent capabilities of the cellular host are also important for the functions of synthetic genetic modules in certain applications aside from the steps of central dogma. For instance, the cellular host must synthesize necessary cofactors for proteins encoded in the implemented synthetic genetic module [54]. Moreover, the selected cell host should exhibit robust adaptation to the specific environmental conditions for the required application. On the other hand, the intrinsic cellular processes of the host should minimally interfere with the function of the implemented synthetic genetic modules [55, 56]. In general, an ideal cellular chassis does not exist in nature because of evolutionary constraints. "A genomic taming" is required for the wild-type cells in order to increase the effectiveness of the implemented synthetic genetic modules.

There are many methods to modify, delete, or insert certain DNA sequences within the genome of cells. CRISPR-Cas systems are by far the most utilized genome modification tools after its repurposing [57]. The CRISPR Cas system was first discovered as an adaptive defense mechanism against bacteriophages in bacteria species [58]. The main effector, Cas ribonuclease protein, binds two different RNAs, one of which contains a unique target nucleotide sequence. Upon recognition of a unique sequence, the effector Cas protein cleaves DNA strands and introduces a double-strand break (DSB) [59]. This DSB can be used to insert and/or remove DNA sequences from the genome of a cellular host [60].

1.3 Sensing modules

Cells continuously monitor their environment via genetically encoded biosensors. The sensing capabilities provide survival advantages to organisms; therefore, organisms have evolved to equip sensors that can detect molecules in their ecological niche. For instance, the urinary tract pathogen *Proteus mirabilis* can detect the presence of urea with a transcriptional factor called UreR and produce urease in the presence of urea [61]. The sensing capabilities of cellular chassis can be augmented synthetically by harnessing natural biosensors found in nature. Mainly, two different types of synthetic biosensors are being used for cell reprogramming which are transcriptional and translational sensing. In this section, two different sensing strategies will be discussed along with works on fine-tuning parameters of genetically encoded biosensors.

1.3.1 Transcriptional sensing strategies

Transcriptional sensing involves using the transcriptional regulation mechanisms to sense signals within cells. TFs play a key role in this process as they bind to specific DNA sequences called promoters in response to environmental stimuli. The expression of target genes is controlled by the binding of TFs to DNA sequences [62]. One famous strategy for coupling environmental stimuli to adaptive responses in bacteria is the two-component signal transduction system. The two-component system is one of the modular signaling systems found in bacteria to help sense and adapt to dynamic changes in environmental conditions and enhance their survival. As the name suggests, the system comprises two parts: a sensor kinase spanning across the membrane and the response regulator located in the cytoplasm [63]. The sensor kinase is a transmembrane protein responsible for sensing external signals such as changes in temperature, pH, available nutrient levels, presence of toxins or antibiotics, and redox state. Following sensing the signal, the protein would undergo autophosphorylation, which in turn activates and causes a conformational change in the second component of the system called the response regulator, which can alter the gene expression by binding to the target DNA region adjacent to a promoter. These genes are crucial genes that take a role in bacterial processes such as virulence, biofilm formation, and antibiotic resistance. Two-component biosensors are engineered to enhance the sensor's recognition specificity. One strategy is to replace the sensor kinase's inducer binding domain with stronger binding. Ma et al. [64] built a red-light controllable system using the light-sensing domain of Cph1 from *Cyanobacteria synechocystis* instead of the original histidine kinase in the system. Similarly, in another study, a fusion of the sensor domains of *Pseudomonas putida* chemoreceptors to the signaling domains of the *E. coli* NarX/NarQ nitrate sensors resulted in an enhanced aromatic acid responsive sensor system [65]. Some of the very old and known examples of the context of transcriptional sensing systems are the lac operon and arabinose operon. The lac operon in *E. coli* is a classic model system for understanding gene regulation consisting of genes involved in lactose metabolism and is controlled by the lac repressor and a catabolite activator protein responding to lactose levels [66]. Similarly, the arabinose operon in *E. coli* regulates genes responsible for arabinose metabolism in response to the presence of arabinose [67].

Synthetic TFs, such as zinc finger proteins, transcriptional activator-like effectors, and Cas9-deficient proteins, are sophisticated methods that allow precise control over gene expression by engineering cells at the transcriptional level [68]. Engineered TFs here are a powerful tool for manipulating cellular behavior as they can interact with specific gene promoters to activate or repress gene transcription [69]. Also, even in the regions where the chromatin is closed, a combination of synthetic TFs is capable of synergistically activating human genes by targeting endogenous gene promoters. This proves the TFs' complex and tunable engineering ability in gene regulation. Also, in 2014, Skjoedt et al. [70] showed how the engineering of prokaryotic transcriptional activators as metabolite biosensors in yeast proves the versatility of transcriptional regulators in reprogramming cellular responses. As CRISPR-based technologies have emerged, it led to the development of Cas9 transcriptional activators that enable precise genome regulation in human cells by linking transcriptional activation domains to the Cas9 protein and guide RNA molecules [10, 71]. Moreover, the reprogramming of gene transcription in industrial yeast strains by engineering TFs, such as the TATA-binding protein, has highlighted the impact of transcriptional engineering on generating desired cellular phenotypes for industrial applications [72]. Similarly, in a 2020 study by Meško et al. [73], tight regulation of gene expression in mammalian cells by engineering a calcium-dependent signaling pathway offered a promising and versatile platform for therapeutic and diagnostic applications. As scientists regulate transcriptional sensing by manipulating gene expression, reshaping cellular functions, and tailoring specific cellular reactions, they use synthetic biology tools, such as synthetic TFs, CRISPR technologies, and customized signaling pathways, which offer advanced cellular manipulation.

1.3.2 Translational sensing strategies

Translation initiation starts at the ribosome binding site (RBS) in bacteria. RBS-based sensing elements are the key components to focus on when engineering cells at the translational level respond to specific ligands while modulating translation. Hence, manipulations of the RBS directly impact protein expression levels, influencing the dynamic range of biosensors. Along with protein expression levels, the translation rates at the cell, protein folding, and the expression of corresponding TFs and reporter genes are mainly affected within the biosensor [74]. In their 2023 study, Wang et al. [75] came up with an erythromycin biosensor with modulated sensitivity and dynamic range by finely adjusting the regulator expression levels in their design by RBS engineering. The concept of RBS engineering has been extended to develop whole-cell biosensors by integrating RBS engineering with the expression of specific genes in a study of the design of a biosensor for monitoring shikimic acid production in *Corynebacterium glutamicum* [76]. Such biosensors have various applications, such as screening and characterizing the "producer strain" for high-yield production [77]. Overall, RBS engineering is a crucial tool in biosensor development, as it helps in designing sensors with enhanced sensitivity, specificity, and dynamic ranges to meet a set of application requirements.

Riboswitches and aptamers are RNA-based sensing elements that can regulate translation by binding to specific ligands [78]. First discovered in bacteria back in 2002, Riboswitches have been used extensively in biosensors. These RNA-based intracellular sensors can bind small metabolites and regulate gene expression and translation. However, some riboswitches are rather complex, bearing protein factors in their structural and functional domains called regulatory RNA units. They consist of an aptamer domain for ligand binding and an expression platform domain for regulating translation in response to structural changes in the aptamer domain [79].

Moreover, aptamers are defined as single-stranded RNA or DNA molecules that can bind to specific targets with high affinity and specificity. They can directly affect gene expression by inducing a conformational change upon binding to their ligands, modulating translation [80]. There are many biosensing applications utilizing aptamers. One of these studies was conducted in 2019, focusing on developing an aptamer-field-effect transistor sensor for detecting phenylalanine to investigate hyperphenylalaninemia models [81]. Translational sensing offers precise and adjustable control of gene expression in response to specific inputs, which is advantageous for synthetic biology. Researchers may create genetic circuits that react to various ligands or environmental stimuli by employing RNA-based sensing components like aptamers and riboswitches. This makes creating biosensors, genetic switches, and other artificial biological devices possible. To sum up, RNA-based components-mediated translational sensing systems provide a flexible method of adjusting gene expression in reaction to certain stimuli. Direct translation regulation is made possible by riboswitches and aptamers, which benefit synthetic biology applications by creating complex genomic circuits with fine-grained control over gene expression.

1.3.3 Synthetic modular biosensor design

Modular design principles guide synthetic biologists when building sensing modules. Following a bottom-up approach, individual sensing elements must be modularized and standardized before being integrated into larger systems. This bottom-up approach would outline an outline for creating versatile and scalable sensing platforms that can be easily modified and interchanged, enhancing their interoperability. One crucial feature of modular sensing platforms is that they are composed of dynamic, adaptable, and programmable sensors. In their 2019 work focusing on designing artificial systems with programmable properties, Rosetti et al. discuss the development of sensors that leverage modular designs comprising dynamic aptamer-based units and synthetic RNA nanodevices that can perform target-responsive regulation of gene expression. This work lays out the foundation of flexibility and adaptability of modular sensing elements [82]. A synthetic receptor platform called EMeRALD (engineered modularized receptors activated via ligand-induced dimerization) developed by Cheng et al. showcases the integration of various sensing elements into biosensors to enhance the scalability and modifiability of the system. This sensing platform is an example of modular assembly of synthetic sensing modules onto a generic signaling scaffold in *E. coli* [83]. This modularity facilitates the integration of different sensor technologies and enhances the system's interoperability and adaptability to various sensing requirements. In summary, modular design principles allow standardizing and modularizing the individual sensing elements, which are integrated into larger genetic circuits to build scalable, easily modifiable, and interoperable platforms.

1.3.4 Optimization of sensing dynamics

In order to optimize sensing dynamics, the response should be fine-tuned at parameters such as promoter strength, TF affinity, and feedback regulation. Such parameters play vital roles in achieving the desired sensing dynamics including response time, sensitivity, and dynamic range [84]. Promoter strength is crucial as it can be optimized to modulate gene expression levels. Different promoters are known to have different strengths. Governed by the nucleic acid sequence, the promoter strength controls the gene expression rate, thereby influencing the dynamic response of the sensing module [85]. A further refinement of sensing is obtained by tuning the TF affinity. TF affinity for specific DNA sequences can impact the binding specificity of the sensing modules [86]. Feedback regulation is yet another great tool to optimize sensing dynamics. Feedback loops can be integrated within the genetic circuits to allow cells to adjust their response to environmental signals dynamically, improving the sensitivity and dynamic range. Feedback regulation serves multiple purposes; it enhances adaptability to external variations like pH, temperature, and oxygen levels, allowing adjustments based on the current environment. It also helps manage internal variability arising from differences between cells and enhances performance by reducing response times, which is vital for biosensing applications. Examples further illustrate the importance of dynamic optimization in synthetic biology. For instance, the use of quorum sensing (QS) systems that naturally allow cells to regulate gene expression in response to changes in cell population through chemical signals produced called autoinducers, coupled with transcriptional factor-based biosensors, has enabled metabolic engineers to allocate carbon flux dynamically, optimizing cell metabolism based on sensing cell population and product levels [87]. To sum up, promoter potency, TF binding, and feedback control allow researchers to enhance sensing dynamics. Engineering these factors leads to building advanced genetic circuits with increased sensitivity, faster response times, and broader dynamic ranges, which opens avenues for innovative uses in metabolic engineering, biosensing, and therapeutic interventions.

1.3.5 Engineering sensitivity and specificity

Optimizations of sensitivity and specificity of sensing modules of engineered cells allow the generation of accurate responses. Improvement of discrimination between different signals, reduction of crosstalk, and background noise are some examples of optimizations synthetic biologists need to focus on. The introduction of negative feedback loops in the genetic circuits of engineered cells is one way to enhance sensitivity and specificity [88]. Employing a negative feedback loop will improve the performance and robustness of the system by reducing the output variation, which helps to minimize the noise by overcoming the dynamic fluctuations of copy number and activity of gene expression. In addition to negative feedback loops, using orthogonal signaling pathways helps isolate the aimed signal from the background, improving the sensing module's specificity [89]. Another critical issue when optimizing sensing is to reduce the crosstalk between multiple signaling pathways. The key to success is carefully designing and characterizing the sensing modules to demolish unwanted, non-specific interactions in various ways [90]. Flachbart et al. constructed a robust transcriptional trans-cinnamic acid responsive biosensor. In this study, they screened variants through fluorescent-activated cell sorting to decrease crosstalk between producing and nonproducing *E. coli* cells. This strategy is an example of improving overall sensor performance by ensuring receiving response to desired input only. Improving the discrimination of closely related signals, which involves fine-tuning the sensitivity of the sensor to detect slight differences in input signals, is also crucial. To tailor the sensor response to distinguish between closely related analytes, the sensing units' detection threshold and dynamic ranges should be adjusted. Such optimizations would enhance the specificity of the overall sensor by providing precise detection and differentiating between multiple signals [91]. To validate the performance of the sensor modules, experiments should be carried out in controlled settings that mimic the desired conditions to confirm the functionality and reliability of the sensor in real-world applications. By applying modern engineering techniques, experimental validation, and rational design concepts, scientists can create sensors that exhibit improved performance by lowering background noise, eliminating crosstalk, and enhancing signal discrimination. These are all necessary for optimal sensitivity and specificity in designed cells.

1.4 Signal processing modules

Synthetic sensing modules are needed to activate intracellular signaling pathways in the presence of external target stimuli. For the engineered cell to decide how to respond to this stimulus, it also needs modules for signal processing the information and regulating the output actuation. Processing may not be required for simpler designs. However, a processing layer is also needed for decision-making before acting on the output for complex genetic circuits [92]. Signal processing modules for synthetic biology have been inspired by unique molecular biological processes. Developing systems that respond to specific signals can be made easier by understanding cellular signaling pathways. Bacterial motility provides guidance for the engineering of directed movement in synthetic systems, whereas biochemical switches and oscillators provide dynamic regulation of gene expression. Memory circuits store and retrieve information for temporal regulation and QS mechanisms promote cooperation among the population. Further growing applications are sequential control and complex decision-making made possible by state machine systems and logic operations. These molecular elements are integrated and optimized according to evolutionary principles, which guarantee that favorable traits are refined by selection pressures. Using synthetic biology, scientists want to develop intricate, adaptable biological systems with accurate actuation and control. In this sense, the essential value of processing modules lies in their capacity to precisely regulate and modify biological systems for various applications, from environmental remediation to targeted drug delivery.

1.4.1 Memory circuits

A genetic or biochemical system that is designed to perform a certain function, such as signal processing or regulating gene expression, is commonly referred to as a "circuit" in synthetic biology. Like electronic circuits, these genetic circuits are often composed of interconnected components such as promoters, repressors, and reporters, which work together to achieve the desired outcome. One specific type of artificial biological circuit designed to store and recall information for a long time is called a "memory circuit" or "memory device." Feedback loops and other techniques are used by memory circuits to maintain a stable state or recall previous inputs. This enables the system to remember previous situations or occurrences and respond properly going forward [93]. In the design of complex gene circuits, transforming momentary data into enduring responses holds significant importance since this feature is common to many biological systems [94]. These natural biological systems include bacteriophage lambda switch, cell division, and differentiation, or Cyanobacteria circadian oscillator. In 2000, two initial examples of synthetic biological circuits were described in *Nature*, which is the world's leading multidisciplinary science journal. Gardner et al. [96] have shown a bistable synthetic gene-regulatory network or toggle switch in *E. coli*. The reciprocal inhibitory pattern of the repressor genes gives rise to the bistability of the toggle. When the inducer is absent, two stable states are possible: one in which promoter 1 starts the transcription of repressor two and another in which promoter 2 starts the transcription of repressor 1. By briefly delivering an inducer unique to the repressor that is currently active, the switch is made. The counteractive repressor's maximal transcription is aided by this inducer until it successfully represses the initially active promoter, maintaining the situation. The switch is activated when heat is applied to the bacterial culture; when IPTG is introduced, it becomes deactive. Green fluorescent protein was used by the researchers' system as an indication [95, 96]. However, Elowitz and Leibler [97] showed how three repressor genes interacted to create a negative feedback loop known as the "repressilator," which caused constant fluctuations in the levels of protein in *E. coli*. The system was producing green fluorescent protein intermittently to indicate its state within individual cells. The resulting oscillations, which usually continued for hours, were happening more slowly than the cell-division cycle. Therefore, the oscillator's state must be passed on through generations. However, this artificial clock produced background noise, which could be explained by random oscillations in its component parts [97]. These early efforts were mainly concerned with building compact genetic circuits with a small number of components. However, current research efforts have broadened our capabilities to encompass substantial multi-gene constructs and even entire genomes.

Synthetic memory circuits exhibit distinct characteristics based on their need for active cellular processes to retain information. Volatile memory circuits, such as transcription-based devices, necessitate ongoing cellular activities to maintain their state, while nonvolatile memory circuits, often reliant on recombination, do not require such continual cellular processes. An essential attribute of volatile memory circuits is their bistability, which typically exists in one of the two stable states with infrequent stochastic transitions between them. Moreover, volatile and nonvolatile memory circuits can undergo reversible or irreversible changes in their state. Following the addition of a signal in irreversible memory, the cell's state changes and remains fixed, whereas in reversible memory, the cell's state changes after the first signal addition and can be reverted to its initial state with the addition of a second signal [95].

Another approach to defining the characteristics of a persistent memory capability lies in between two primary gene network structures: the double-negative feedback loops and the positive feedback loop. These approaches have undergone exploration within various contexts, each carrying its own set of strengths and limitations. The former presents a simpler circuit requiring fewer components for engineering, while the latter offers greater flexibility for modification and fine-tuning due to its heightened complexity. Gardner et al. [96] previously described that one of the initial examples of synthetic circuits is an illustration of a double-negative feedback loop (Figure 1.2A), while the first positive-feedback loop, which consistently and dependably conveyed memory across numerous cell divisions, was constructed within *S. cerevisiae* by Ajo-Franklin et al. [98]. They showcased a memory system composed of artificial TFs arranged in an autoregulatory positive feedback pattern, preserving an induced state in an inheritable manner following a temporary stimulus. Until this study, since 2001, autoregulatory positive feedback designs have been demonstrated multiple times by other research groups; however, no synthetic circuit study conducted in eukaryotes has proven a predictable and reliable circuit behavior [95, 98, 99] (Figure 1.2B).

1.4.2 Quorum sensing (QS)

One more biological system that was envisaged for reengineering to construct synthetic gene circuits was QS. Bacterial populations use QS as a critical mechanism to communicate within their own species or between different species and to organize coordinated behavior. In recent years, researchers in the field of synthetic biology have appreciated

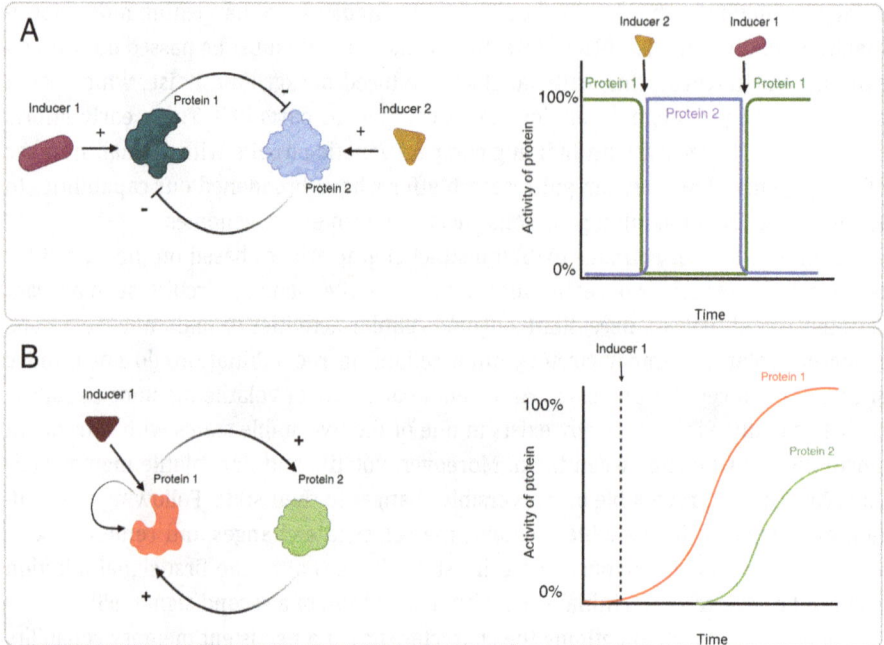

Figure 1.2: Genetic feedback loops for memory formation. (A) Example illustration of double-negative feedback loop memory circuit. Inducer1 promotes the production of Protein1, while Protein1 inhibits the production of Protein2. Similarly, Inducer2 promotes the production of Protein2, while Protein2 inhibits the production of Protein1. Due to this mutual inhibition, the system operates in two stable states: High Protein1 – Low (nearly no) Protein2 or High Protein2 – Low (nearly no) Protein1. Therefore, both Protein1 and Protein2 cannot be produced at high levels at the same time. The system must be in either of two states. (B) Example illustration of positive feedback loop memory circuit. Inducer1 stimulates the production of Protein1, while Protein1 induces the production of Protein2. Protein2 also acts as an inducer for Protein1, which also induces its own production (acting as autoinducer). Through this mechanism, once induced, Protein1 production remains at consistently high levels (created with BioRender.com).

the remarkable properties of QS in creating genetic circuits sensitive to population density. Based on this definition, advances have been made in creating dynamic, adaptive, and occasionally multicellular systems. This has contributed significantly to the bioproduction potential of metabolic engineering and enriched the spatial and temporal complexities in the field of synthetic biology [100]. Quorum-sensing bacteria express and release chemical signaling molecules, also known as autoinducers, which increase concentration depending on cell density. Determination of the minimum stimulatory concentration of these autoinducers leads to changes in gene expression. Gram-negative and gram-positive bacteria use quorum-sensing communication circuits to regulate a variety of physiological activities including competence, virulence, symbiosis, motility, conjugation, spore formation, antibiotic production, and biofilm formation. Gram-positive bacteria generally utilize processed oligo-peptides while gram-negative bacteria utilize acyl-homoserine lactones (AHLs) for communication. These autoinducers in the extracellular space penetrate the cell membrane and directly or indirectly control the expression of specific genes, allowing bacteria to respond to their surrounding medium [101–103].

AHL (AI-1) systems used by gram-negative bacteria stand out among the most widely recognized QS mechanisms. This mechanism was first identified in the marine bacterium *V. fischeri* and was observed to regulate luminescence production. In this natural mechanism, luxI is involved in the synthesis of the AHL molecule, which then freely penetrates the cell membrane, as depicted in Figure 1.3. As cellular density increases, AHL levels simultaneously increase. Once a critical concentration of AHL is reached, AHL binds to luxR, forming a complex that activates the bidirectional lux promoter. Activation of this promoter leads to additional transcription of luxI and luxR as well as transcription of genes such as luciferase (luxCDABE). This creation of a positive feedback loop characterizes many QS systems [104]. Synthetic biology approaches can use autoinducers such as AHL to engineer gene signals that respond to changes in cell density for cell-cell communication. These systems are important as activation modules in synthetic biology because they provide invaluable control over the timing and occurrence of transcription and its synchrony [105].

AI-2 (autoinducer 2) is another QS mechanism. More than 55 bacteria species, including gram-positive and gram-negative ones, share the AI-2 system. AI-2 is produced by the luxS enzyme [106]. Like the aforementioned AHL system, the autoinducer-2 (AI-2) mechanism in QS contains a positive feedback loop. An increase in AI-2 detection in the system causes an increase in AI-2 production, which in turn establishes a cell density-dependent positive autoregulatory circuit for AI-2 synthesis. AI-2 molecules are considered a potential key for interspecies communication due to their presence in different classes of bacterial populations, while AHL molecules are highly specific within the species that produce them [107, 108].

To adapt these systems for synthetic biology applications, these systems can be manipulated in different ways and strategies. For example, regulatory molecules can be modified to control signal sensitivity, alter signal reception, mutate the promoter to modify the response to the quorum-sensing molecule of interest, or introduce additional genetic circuits into the cell to program the response that this molecule elicits within the cell [104].

1.4.3 Logic operations

Genetic logic gates, as one of the signal processing modules in the field of synthetic biology, are systems that enable the creation of interconnected gene regulatory networks to provide precise control over gene expression and cellular behavior. These gates mimic the working principle of electronic logic gates in a biological setting and are based on the operating principles of the Boolean function [109]. Although many

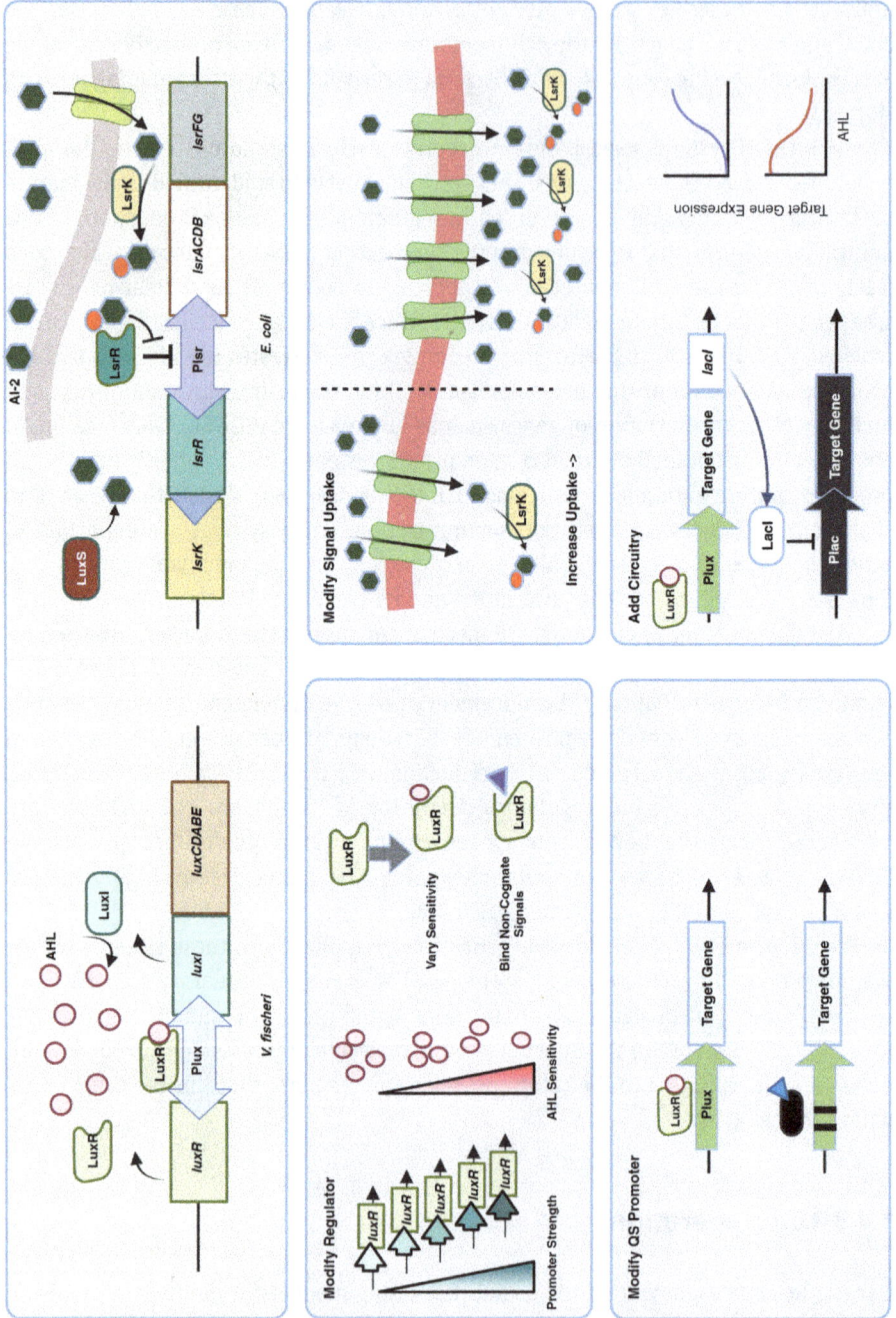

Figure 1.3: Schematic of the reengineering of quorum sensing signals in synthetic biology (adapted from [104]. Created with BioRender.com).

types of genetic logic gates have been developed, AND, OR, and NOT gates are the most common examples of these systems, and each of them interprets genetic information in different ways and produces an output accordingly [110]. They are crucial for designing complex genetic circuits that allow dynamic control of gene expression, cellular state transitions, and memory-based systems. Furthermore, by integrating multiple genetic logic gates into the cell, more complex genetic circuits can be constructed that can perform functions such as multiplexers, half-adders, memory devices, and sequential logic circuits [111]. Logic operation-based systems are phenomenal due to their ability to create artificial circuits in the cell by using natural regulators and make it possible to perform and control certain tasks that the cell cannot perform under natural conditions.

1.4.3.1 AND gate, OR gate, and NOT gate

An AND gate (Figure 1.4) is a circuit that requires the presence of two (or more) specific input signals to produce a designated output. It mimics the logical AND function: both conditions must be met simultaneously for the system to activate a downstream response [112]. In synthetic biology, an AND gate is typically used in conjunction with several regulatory elements to regulate the production of a reporter gene or functional protein. These regulatory elements may be DNA-binding proteins, TFs, or promoters as common examples. These gates usually use two promoters (dual promoters), each responding to different inputs, and gene expression occurs only when both promoters are activated [110]. Alternatively, AND gates can include two different TFs, each binding to different sites in the promoter region of the reporter gene, and transcription is initiated only when both TFs are present [113]. Some designs use hybrid promoters that have binding sites for two different regulatory proteins, and only when both proteins bind to the promoter, the RNA polymerase can effectively initiate transcription [114].

An OR gate (Figure 1.4) is a circuit where at least one of the specific input signals produces a designated output. It mimics the logical OR function: any one of the conditions must be met for the system to activate a downstream response [115]. The OR gate can use tandem promoters, each responding to a different input signal, and if one of the promoters gets activated, the downstream gene will be expressed. Alternatively, two separate expression cassettes, each with its own promoter and regulatory elements, can be used, and activation of either cassette leads to the expression of the reporter gene [116]. Some designs use hybrid promoters that have multiple binding sites for different regulatory proteins.

A NOT gate (Figure 1.4) is a circuit that also acts as an inverter and produces an output opposite to the input. In synthetic biology, it ensures no output (OFF state) in the presence of an input signal (ON), reversely, the absence of an input signal (OFF) leads to the expression of a reporter gene or functional protein (ON state). The most common approach used in this system is to use an inducible input promoter to generate a transcriptional repressor. The repressor will then bind to an operator sequence within the cognate promoter and eliminate the output [117].

In synthetic biology, hybrid systems have been developed in which these gates are combined to develop more complex, tightly controllable, nonleaky signal processing modules. In this context, NAND gate represents NOT and AND, which is actually the inverse of AND gate: when all the inputs are present, it gives no output. On the other hand, NOR gate represents NOT and OR, which is the inverse of OR gate: when there is zero input, it produces the output signal [115]. XOR gate (exclusive OR) is another common type of logic operation used in synthetic biology, which produces the output only when exactly one input is present, but not both [110].

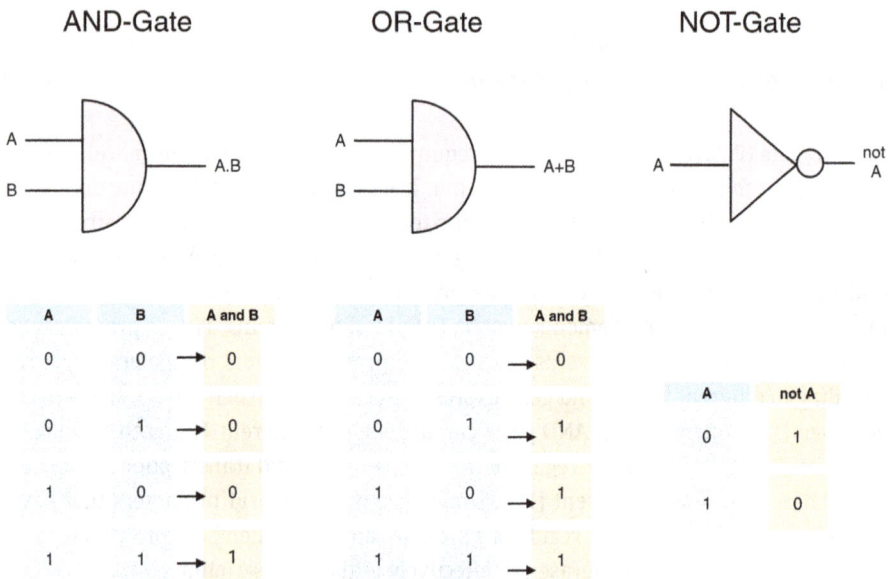

Figure 1.4: Illustration depicting the summary of logic gate truth tables (created with BioRender.com).

1.4.3.2 Proteins for logical operations

dCas9 (dead-Cas9) is an artificial enzyme that is the catalytically inactive version of Cas9, binding to DNA but not cutting it. It can be combined with other effector domains to regulate gene expression without altering the DNA sequence. It can also be used to repress or activate gene transcription by either blocking or recruiting TFs. Selective regulation of gene expression allows the creation of NOT, NOR, and NAND gates [118]. For example, in logic gate circuit using Cas9, two different promoters constitute the inducers of the system. Promoter-1 is responsible for Cas9 transcription, while promoter-2 is responsible for the generation of target-of-interest-specific sgRNA.

Only when both transcription events occur, Cas9-sgRNA complex come together to cut off the gene that is responsible for the transcription of a repressor molecule. Since the repressor molecule cannot be transcribed anymore, it cannot repress the transcription of the effector, which creates the output signal. This is a common AND gate operation that is built using CRISPR/Cas9 system [119].

Recombinases, or DNA recombinases, are a group of enzymes that can target and modify specific sequences in DNA. Originally, their key role was to perform the homologous recombination event, which is important in DNA-repair mechanisms and crossing-over [120]. While constructing logic gates in synthetic biology, these site-specific recombinases are used to integrate, delete, or invert DNA sequences [121]. Regulation and decision of these distinct activity types depend on specific nucleotide sequences (or recognition sites). Since recombinase enzymes recognize highly specific DNA sequences, using them to have precise control over gene expression for constructing genetic logical operations is feasible [122]. For example, *loxP*-flanking and/or *frt*-flanking genes can be placed into a host cell. In this way, which gene is in the on/off state can be tightly controlled. To illustrate, if Cre recombinase is present in the system as an input, the *loxP*-flanking gene will be rearranged. Similarly, if Flp recombinase is present as an input, the *frt*-flanking gene will be rearranged. The mechanism of action in this arrangement may vary depending on how the system is designed. For example, when *frt* or *loxP* regions are placed in the same orientation, the Flp and Cre enzymes perform deletion (excision), while when they are placed in the opposite orientation, they perform inversion (flipping). Logic gates with two or more inputs can be created because the systems can be utilized in conjunction with one another and controlled at multiple points. In this regard, the two-input system in the example can be used to produce a logic gate, and a mechanism that can accept four distinct outputs can be established [123–125]. In addition, creating logic gates with memory using irreversible recombination is possible. For instance, an integrated DNA fragment stays in place when irreversible recombination occurs, offering a stable event memory. This enables the genetic circuit to "remember" previous inputs [126].

1.5 Conclusion

Synthetic biology has paved the way to reprogram cells for diverse applications by providing tools for effective genetic engineering and modularizing biological parts and modules. This modularization provides a systemic approach for cell reprogramming in which synthetic genetic modules are built, tested, and optimized within DBTL cycles. In this chapter, we summarized the methodologies that are utilized to construct synthetic genetic modules for cell reprogramming such as de novo DNA synthesis and genome engineering. In addition, commonly used signal sensing and processing modules were presented. Many studies have used these modules for cell reprogramming, which revolutionized diverse areas including medicine [127, 128], agriculture [129–131], and manufacturing [132, 133]. However, there are numerous efforts required to realize the full potential of cell reprogramming.

The main hurdles faced in designing sensing modules for whole-cell biosensors are reliability, scalability, and robustness. Providing consistent performance across changing conditions with high specificity and sensitivity is crucial for sensing modules [134]. As computational tools and modeling are growing, the sensor behavior and performance can now be predicted before development, which eases the rational design of the sensors [135]. Furthermore, the reliability of sensing modules can be increased by tightly controlling the output generation with additional translation regulators, thereby decreasing inherent leakiness [136]. The scalability of biosensing can be improved by distributing sensing modules among cells, which can be wired via communication circuits such as QS [137]. Robust sensing modules can be achieved by using synthetic noise-canceling genetic circuits [138]. As seen, the growing tools of synthetic biology and outbox genetic circuit design can provide means to improve the existing sensing modules.

Current designs of signal processing modules are impeded by cellular burden, a consequence of the requirement for multiple expression cassettes. For example, a recombinase-based AND gate necessitates the expression of two recombinase proteins in addition to the regulatory TFs that control their production [126]. This cellular burden significantly limits the scalability of synthetic signal processing modules and actuated outputs. To address this issue, RNA-based signal processing modules can be employed to reduce cellular burden. For instance, multiple orthogonal trigger-toehold RNA switches can be engineered to construct logic gates [139]. Furthermore, RNA-based signal processing modules facilitate the implementation of complex operations due to the predictable nature of RNA-RNA and RNA-DNA interactions, which can be readily calculated in silico [140]. Additionally, these interactions can be manipulated using proteins, such as Cas, enhancing the versatility and functionality of RNA-based systems [141].

Although only a few examples are provided to illustrate efforts aimed at enhancing the capabilities of existing synthetic modules for cell reprogramming, there is substantial interest within the scientific community in designing novel genetic modules and functions for diverse applications. Earth's biodiversity encompasses a vast array of intriguing molecular machines, many of which remain undiscovered and not fully understood, but these natural systems, when repurposed, might offer a way by expanding the functional repertoire of biological parts to design synthetic cells with unique functions.

References

[1]   Konieczny L, Roterman-Konieczna I, Spólnik P. Regulation in Biological Systems. Systems Biology: Functional Strategies of Living Organisms [Internet]. 2nd edition, Springer; 2023. https://doi.org/ 10.1007/978-3-031-31557-2_4.

[2]   Brown TA. How Genomes Evolve. Genomes. 2nd edition, Wiley-Liss; 2002.

[3]   De Rodas B, Youmans BP, Danzeisen JL, Tran H, Johnson TJ. Microbiome profiling of commercial pigs from farrow to finish. J Anim Sci 2018;96:1778–94. https://doi.org/10.1093/jas/sky109.

[4]   Liu Y-H, Mohamad OAA, Gao L, Xie Y-G, Abdugheni R, Huang Y, et al. Sediment prokaryotic microbial community and potential biogeochemical cycle from saline lakes shaped by habitat. Microbiol Res 2023;270:127342. https://doi.org/10.1016/j.micres.2023.127342.

[5]   Nicolosi G, Gonzalez-Pimentel JL, Piano E, Isaia M, Miller AZ. First insights into the bacterial diversity of Mount Etna volcanic caves. Microb Ecol 2023;86:1632–45. https://doi.org/10.1007/s00248-023-02181-2.

[6]   Pati S, Banerjee S, Sengupta A, Sarma J, Shaheen S, Tenguria S, et al. Chapter 16 – Adaptation Strategies of Thermophilic Microbes. In: Kumar A, Tenguria S, editors. Bacterial Survival in the Hostile Environment, Academic Press; 2023, pp. 231–49. https://doi.org/10.1016/B978-0-323-91806-0.00012-6.

[7]   Ahmed Z, Zulfiqar H, Khan AA, Gul I, Dao F-Y, Zhang Z-Y, et al. iThermo: A sequence-based model for identifying thermophilic proteins using a multi-feature fusion strategy. Front Microbiol 2022;13. https://doi.org/10.3389/fmicb.2022.790063.

[8]   Goldstein RA. Amino-acid interactions in psychrophiles, mesophiles, thermophiles, and hyperthermophiles: Insights from the quasi-chemical approximation. Protein Sci 2007;16:1887–95. https://doi.org/10.1110/ps.072947007.

[9]   Bala S, Garg D, Thirumalesh BV, Sharma M, Sridhar K, Inbaraj BS, et al. Recent strategies for bioremediation of emerging pollutants: A review for a green and sustainable environment. Toxics 2022;10:484. https://doi.org/10.3390/toxics10080484.

[10]   Atuchin VV, Asyakina LK, Serazetdinova YR, Frolova AS, Velichkovich NS, Prosekov AY. Microorganisms for bioremediation of soils contaminated with heavy metals. Microorganisms 2023;11:864. https://doi.org/10.3390/microorganisms11040864.

[11]   Kemter AM, Patry RT, Arnold J, Hesser LA, Campbell E, Ionescu E, et al. Commensal bacteria signal through TLR5 and AhR to improve barrier integrity and prevent allergic responses to food. Cell Rep 2023;42:113153. https://doi.org/10.1016/j.celrep.2023.113153.

[12]   Alam K, Mazumder A, Sikdar S, Zhao Y-M, Hao J, Song C, et al. Streptomyces: The biofactory of secondary metabolites. Front Microbiol 2022;13:968053. https://doi.org/10.3389/fmicb.2022.968053.

[13]   Li C, Zhang R, Wang J, Wilson LM, Yan Y. Protein engineering for improving and diversifying natural products biosynthesis. Trends Biotechnol 2020;38:729–44. https://doi.org/10.1016/j.tibtech.2019.12.008.

[14]   Hadadi N, Hatzimanikatis V. Design of computational retrobiosynthesis tools for the design of *de novo* synthetic pathways. Curr Opin Chem Biol 2015;28:99–104. https://doi.org/10.1016/j.cbpa.2015.06.025.

[15]   Lin G-M, Warden-Rothman R, Voigt CA. Retrosynthetic design of metabolic pathways to chemicals not found in nature. Curr Opin Syst Biol 2019;14:82–107. https://doi.org/10.1016/j.coisb.2019.04.004.

[16]   Siew YY, Zhang W. Downstream processing of recombinant human insulin and its analogues production from *E. coli* inclusion bodies. Bioresour Bioprocess 2021;8:65. https://doi.org/10.1186/s40643-021-00419-w.

[17]   Voigt CA, Keasling JD. Programming cellular function. Nat Chem Biol 2005;1:304–07. https://doi.org/10.1038/nchembio1105-304.

[18] Church GM, Elowitz MB, Smolke CD, Voigt CA, Weiss R. Realizing the potential of synthetic biology. Nat Rev Mol Cell Biol 2014;15:289–94. https://doi.org/10.1038/nrm3767.

[19] Voigt CA. Genetic parts to program bacteria. Curr Opin Biotech 2006;17:548–57. https://doi.org/10.1016/j.copbio.2006.09.001.

[20] Khalil AS, Collins JJ. Synthetic biology: applications come of age. Nat Rev Genet 2010;11:367–79. https://doi.org/10.1038/nrg2775.

[21] Kitano S, Lin C, Foo JL, Chang MW. Synthetic biology: learning the way toward high-precision biological design. PLoS Biol 2023;21:e3002116. https://doi.org/10.1371/journal.pbio.3002116.

[22] Ma S, Tang N, Tian J. DNA synthesis, assembly and applications in synthetic biology. Curr Opin Chem Biol 2012;16:260–67. https://doi.org/10.1016/j.cbpa.2012.05.001.

[23] Hoose A, Vellacott R, Storch M, Freemont PS, Ryadnov MG. DNA synthesis technologies to close the gene writing gap. Nat Rev Chem 2023;7:144–61. https://doi.org/10.1038/s41570-022-00456-9.

[24] Cleary MA, Kilian K, Wang Y, Bradshaw J, Cavet G, Ge W, et al. Production of complex nucleic acid libraries using highly parallel in situ oligonucleotide synthesis. Nat Methods 2004;1:241–48. https://doi.org/10.1038/nmeth724.

[25] Kosuri S, Eroshenko N, LeProust EM, Super M, Way J, Li JB, et al. Scalable gene synthesis by selective amplification of DNA pools from high-fidelity microchips. Nat Biotechnol 2010;28:1295–99. https://doi.org/10.1038/nbt.1716.

[26] Stemmer WPC, Crameri A, Ha KD, Brennan TM, Heyneker HL. Single-step assembly of a gene and entire plasmid from large numbers of oligodeoxyribonucleotides. Gene 1995;164:49–53. https://doi.org/10.1016/0378-1119(95)00511-4.

[27] Gibson DG, Young L, Chuang R-Y, Venter JC, Hutchison CA, Smith HO. Enzymatic assembly of DNA molecules up to several hundred kilobases. Nat Methods 2009;6:343–45. https://doi.org/10.1038/nmeth.1318.

[28] Tolmachov O. Designing plasmid vectors. In: Walther W, Stein US, editors. Gene Therapy of Cancer: Methods and Protocols, Totowa, NJ: Humana Press; 2009, pp. 117–29. https://doi.org/10.1007/978-1-59745-561-9_6.

[29] Bolivar F, Rodriguez RL, Betlach MC, Boyer HW. Construction and characterization of new cloning vehicles I. Ampicillin-resistant derivatives of the plasmid pMB9. Gene 1977;2:75–93. https://doi.org/10.1016/0378-1119(77)90074-9.

[30] Bolivar F, Rodriguez RL, Greene PJ, Betlach MC, Heyneker HL, Boyer HW, et al. Construction and characterization of new cloning vehicle. II. A multipurpose cloning system. Gene 1977;2:95–113. https://doi.org/10.1016/0378-1119(77)90000-2.

[31] Vieira J, Messing J. The pUC plasmids, an M13mp7-derived system for insertion mutagenesis and sequencing with synthetic universal primers. Gene 1982;19:259–68. https://doi.org/10.1016/0378-1119(82)90015-4.

[32] Li X, Xie Y, Liu M, Tai C, Sun J, Deng Z, et al. oriTfinder: A web-based tool for the identification of origin of transfers in DNA sequences of bacterial mobile genetic elements. Nucleic Acids Res 2018;46:W229–34. https://doi.org/10.1093/nar/gky352.

[33] Wang Z, Jin L, Yuan Z, Węgrzyn G, Węgrzyn A. Classification of plasmid vectors using replication origin, selection marker and promoter as criteria. Plasmid 2009;61:47–51. https://doi.org/10.1016/j.plasmid.2008.09.003.

[34] Lutz R, Bujard H. Independent and tight regulation of transcriptional units in *Escherichia coli* via the LacR/O, the TetR/O and AraC/I1-I2 regulatory elements. Nucleic Acids Res 1997;25:1203–10. https://doi.org/10.1093/nar/25.6.1203.

[35] Shetty RP, Endy D, Knight TF. Engineering BioBrick vectors from BioBrick parts. J Biol Eng 2008;2:5. https://doi.org/10.1186/1754-1611-2-5.

[36] Silva-Rocha R, Martínez-García E, Calles B, Chavarría M, Arce-Rodríguez A, de las Heras A, et al. The Standard European Vector Architecture (SEVA): A coherent platform for the analysis and

deployment of complex prokaryotic phenotypes. Nucleic Acids Res 2013;41:D666–75. https://doi. org/10.1093/nar/gks1119.

[37]   Heap JT, Pennington OJ, Cartman ST, Minton NP. A modular system for *Clostridium* shuttle plasmids. J Microbiol Methods 2009;78:79–85. https://doi.org/10.1016/j.mimet.2009.05.004.

[38]   Bird JE, Marles-Wright J, Giachino A. A user's guide to golden gate cloning methods and standards. ACS Synth Biol 2022;11:3551–63. https://doi.org/10.1021/acssynbio.2c00355.

[39]   Rajpathak S, Vyawahare R, Patil N, Sivaram A. Fundamental techniques of recombinant DNA transfer. In: Patil N, Sivaram A, editors. A Complete Guide to Gene Cloning: From Basic to Advanced, Cham: Springer International Publishing; 2022, pp. 79–95. https://doi.org/10.1007/978-3-030-96851-9_6.

[40]   Neumann E, Schaefer-Ridder M, Wang Y, Hofschneider PH. Gene transfer into mouse lyoma cells by electroporation in high electric fields. EMBO J 1982;1:841–45.

[41]   Villemejane J, Mir LM. Physical methods of nucleic acid transfer: General concepts and applications. Br J Pharmacol 2009;157:207–19. https://doi.org/10.1111/j.1476-5381.2009.00032.x.

[42]   Du X, Wang J, Zhou Q, Zhang L, Wang S, Zhang Z, et al. Advanced physical techniques for gene delivery based on membrane perforation. Drug Deliv 2018;25:1516–25. https://doi.org/10.1080/10717544.2018.1480674.

[43]   Capecchi MR. High efficiency transformation by direct microinjection of DNA into cultured mammalian cells. Cell 1980;22:479–88. https://doi.org/10.1016/0092-8674(80)90358-X.

[44]   Klein TM, Wolf ED, Wu R, Sanford JC. High-velocity microprojectiles for delivering nucleic acids into living cells. Nature 1987;327:70–73. https://doi.org/10.1038/327070a0.

[45]   Miller DL, Song J. Tumor growth reduction and DNA transfer by cavitation-enhanced high-intensity focused ultrasound *in vivo*. Ultrasound Med Biol 2003;29:887–93. https://doi.org/10.1016/S0301-5629(03)00031-0.

[46]   Tsukakoshi M, Kurata S, Nomiya Y, Ikawa Y, Kasuya T. A novel method of DNA transfection by laser microbeam cell surgery. Appl Phys B 1984;35:135–40. https://doi.org/10.1007/BF00697702.

[47]   Plank C, Schillinger U, Scherer F, Bergemann C, Rémy J-S, Krötz F, et al. The magnetofection method: using magnetic force to enhance gene delivery. 2003;384:737–47. https://doi.org/10.1515/BC.2003.082.

[48]   Calcium phosphate–mediated transfection of eukaryotic cells. Nat Methods 2005;2:319–20. https://doi.org/10.1038/nmeth0405-319.

[49]   van den Berg AIS, Yun C-O, Schiffelers RM, Hennink WE. Polymeric delivery systems for nucleic acid therapeutics: Approaching the clinic. J Control Release 2021;331:121–41. https://doi.org/10.1016/j.jconrel.2021.01.014.

[50]   Mehta M, Bui TA, Yang X, Aksoy Y, Goldys EM, Deng W. Lipid-based nanoparticles for drug/gene delivery: An overview of the production techniques and difficulties encountered in their industrial development. ACS Mater Au 2023;3:600–19. https://doi.org/10.1021/acsmaterialsau.3c00032.

[51]   Bulcha JT, Wang Y, Ma H, Tai PWL, Gao G. Viral vector platforms within the gene therapy landscape. Sig Transduct Target Ther 2021;6:1–24. https://doi.org/10.1038/s41392-021-00487-6.

[52]   Harris HMB, Hill C. A place for viruses on the tree of life. Front Microbiol 2021;11. https://doi.org/10.3389/fmicb.2020.604048.

[53]   Cabezón E, Ripoll-Rozada J, Peña A, de la Cruz F, Arechaga I. Towards an integrated model of bacterial conjugation. FEMS Microbiol Rev 2015;39:81–95. https://doi.org/10.1111/1574-6976.12085.

[54]   Ong NT, Olson EJ, Tabor JJ. Engineering an *E. coli* near-infrared light sensor. ACS Synth Biol 2018;7:240–48. https://doi.org/10.1021/acssynbio.7b00289.

[55]   Stone A, Youssef A, Rijal S, Zhang R, Tian X-J. Context-dependent redesign of robust synthetic gene circuits. Trends Biotechnol 2024;0. https://doi.org/10.1016/j.tibtech.2024.01.003.

[56]   Costello A, Badran AH. Synthetic biological circuits within an orthogonal central dogma. Trends Biotechnol 2021;39:59–71. https://doi.org/10.1016/j.tibtech.2020.05.013.

[57] Cong L, Ran FA, Cox D, Lin S, Barretto R, Habib N, et al. Multiplex genome engineering using CRISPR/Cas systems. Science 2013;339:819–23. https://doi.org/10.1126/science.1231143.

[58] Sapranauskas R, Gasiunas G, Fremaux C, Barrangou R, Horvath P, Siksnys V. The *Streptococcus thermophilus* CRISPR/Cas system provides immunity in *Escherichia coli*. Nucleic Acids Res 2011;39:9275–82. https://doi.org/10.1093/nar/gkr606.

[59] Jinek M, Chylinski K, Fonfara I, Hauer M, Doudna JA, Charpentier E. A programmable dual-RNA–guided DNA endonuclease in adaptive bacterial immunity. Science 2012;337:816–21. https://doi.org/10.1126/science.1225829.

[60] Anzalone AV, Koblan LW, Liu DR. Genome editing with CRISPR–Cas nucleases, base editors, transposases and prime editors. Nat Biotechnol 2020;38:824–44. https://doi.org/10.1038/s41587-020-0561-9.

[61] Köse S, Ahan RE, Köksaldı İÇ, Olgaç A, Kasapkara ÇS, Şeker UÖŞ. Multiplexed cell-based diagnostic devices for detection of renal biomarkers. Biosens Bioelectron 2023;223:115035. https://doi.org/10.1016/j.bios.2022.115035.

[62] Ganesh I, Gwon D, Lee JW. Gas-sensing transcriptional regulators. Biotechnol J 2020;15:1900345. https://doi.org/10.1002/biot.201900345.

[63] West AH, Stock AM. Histidine kinases and response regulator proteins in two-component signaling systems. Trends Biochem Sci 2001;26:369–76. https://doi.org/10.1016/S0968-0004(01)01852-7.

[64] Ma S, Luo S, Wu L, Liang Z, Wu J-R. Re-engineering the two-component systems as light-regulated in *Escherichia coli*. J Biosci 2017;42:565–73. https://doi.org/10.1007/s12038-017-9711-8.

[65] Luu RA, Schomer RA, Brunton CN, Truong R, Ta AP, Tan WA, et al. Hybrid two-component sensors for identification of bacterial chemoreceptor function. Appl Environ Microbiol 2019;85:e01626–19. https://doi.org/10.1128/AEM.01626-19.

[66] Jacob F, Monod J. Genetic regulatory mechanisms in the synthesis of proteins. J Mol Biol 1961;3:318–56. https://doi.org/10.1016/S0022-2836(61)80072-7.

[67] Franklin NC, Yarmolinsky MB. Arabinose operon. Nat New Biol 1971;234:98–98. https://doi.org/10.1038/newbio234098b0.

[68] Gao X, Tsang JCH, Gaba F, Wu D, Lu L, Liu P. Comparison of TALE designer transcription factors and the CRISPR/dCas9 in regulation of gene expression by targeting enhancers. Nucleic Acids Res 2014;42:e155. https://doi.org/10.1093/nar/gku836.

[69] Perez-Pinera P, Kocak DD, Vockley CM, Adler AF, Kabadi AM, Polstein LR, et al. RNA-guided gene activation by CRISPR-Cas9–based transcription factors. Nat Methods 2013;10:973–76. https://doi.org/10.1038/nmeth.2600.

[70] Skjoedt ML, Snoek T, Kildegaard KR, Arsovska D, Eichenberger M, Goedecke TJ, et al. Engineering prokaryotic transcriptional activators as metabolite biosensors in yeast. Nat Chem Biol 2016;12:951–58. https://doi.org/10.1038/nchembio.2177.

[71] Mali P, Esvelt KM, Church GM. Cas9 as a versatile tool for engineering biology. Nat Methods 2013;10:957–63. https://doi.org/10.1038/nmeth.2649.

[72] Hou L, Cao X, Wang C, Lu M. Effect of overexpression of transcription factors on the fermentation properties of *Saccharomyces cerevisiae* industrial strains. Lett Appl Microbiol 2009;49:14–19. https://doi.org/10.1111/j.1472-765X.2009.02615.x.

[73] Meško M, Lebar T, Dekleva P, Jerala R, Benčina M. Engineering and rewiring of a calcium-dependent signaling pathway. ACS Synth Biol 2020;9:2055–65. https://doi.org/10.1021/acssynbio.0c00133.

[74] Salis HM, Mirsky EA, Voigt CA. Automated design of synthetic ribosome binding sites to control protein expression. Nat Biotechnol 2009;27:946–50. https://doi.org/10.1038/nbt.1568.

[75] Wang Y, Li S, Xue N, Wang L, Zhang X, Zhao L, et al. Modulating sensitivity of an erythromycin biosensor for precise high-throughput screening of strains with different characteristics. ACS Synth Biol 2023;12:1761–71. https://doi.org/10.1021/acssynbio.3c00059.

[76] Liu C, Zhang B, Liu Y-M, Yang K-Q, Liu S-J. New intracellular shikimic acid biosensor for monitoring shikimate synthesis in corynebacterium glutamicum. ACS Synth Biol 2018;7:591–601. https://doi.org/10.1021/acssynbio.7b00339.

[77] Siedler S, Khatri NK, Zsohár A, Kjærbølling I, Vogt M, Hammar P, et al. Development of a bacterial biosensor for rapid screening of yeast p-coumaric acid production. ACS Synth Biol 2017;6:1860–69. https://doi.org/10.1021/acssynbio.7b00009.

[78] Mandal M, Breaker RR. Gene regulation by riboswitches. Nat Rev Mol Cell Biol 2004;5:451–63. https://doi.org/10.1038/nrm1403.

[79] Chan CW, Mondragón A. Crystal structure of an atypical cobalamin riboswitch reveals RNA structural adaptability as basis for promiscuous ligand binding. Nucleic Acids Res 2020;48:7569–83. https://doi.org/10.1093/nar/gkaa507.

[80] Ilgu M, Nilsen-Hamilton M. Aptamers in analytics. Analyst 2016;141:1551–68. https://doi.org/10.1039/c5an01824b.

[81] Cheung KM, Yang K-A, Nakatsuka N, Zhao C, Ye M, Jung ME, et al. Phenylalanine monitoring via aptamer-field-effect transistor sensors. ACS Sens 2019;4:3308–17. https://doi.org/10.1021/acssensors.9b01963.

[82] Rossetti M, Del Grosso E, Ranallo S, Mariottini D, Idili A, Bertucci A, et al. Programmable RNA-based systems for sensing and diagnostic applications. Anal Bioanal Chem 2019;411:4293–302. https://doi.org/10.1007/s00216-019-01622-7.

[83] Chang H-J, Zúñiga A, Conejero I, Voyvodic PL, Gracy J, Fajardo-Ruiz E, et al. Programmable receptors enable bacterial biosensors to detect pathological biomarkers in clinical samples. Nat Commun 2021;12:5216. https://doi.org/10.1038/s41467-021-25538-y.

[84] Hansen AS, O'Shea EK. Promoter decoding of transcription factor dynamics involves a trade-off between noise and control of gene expression. Mol Syst Biol 2013;9:704. https://doi.org/10.1038/msb.2013.56.

[85] Li J, Zhang Y. Relationship between promoter sequence and its strength in gene expression. Eur Phys J E 2014;37:86. https://doi.org/10.1140/epje/i2014-14086-1.

[86] Aditham AK, Markin CJ, Mokhtari DA, DelRosso N, Fordyce PM. High-throughput affinity measurements of transcription factor and DNA mutations reveal affinity and specificity determinants. Cell Syst 2021;12:112–27, e11. https://doi.org/10.1016/j.cels.2020.11.012.

[87] Ge C, Yu Z, Sheng H, Shen X, Sun X, Zhang Y, et al. Redesigning regulatory components of quorum-sensing system for diverse metabolic control. Nat Commun 2022;13:2182. https://doi.org/10.1038/s41467-022-29933-x.

[88] Kelly CL, Harris AWK, Steel H, Hancock EJ, Heap JT, Papachristodoulou A. Synthetic negative feedback circuits using engineered small RNAs. Nucleic Acids Res 2018;46:9875–89. https://doi.org/10.1093/nar/gky828.

[89] Daringer NM, Dudek RM, Schwarz KA, Leonard JN. Modular extracellular sensor architecture for engineering mammalian cell-based devices. ACS Synth Biol 2014;3:892–902. https://doi.org/10.1021/sb400128g.

[90] Flachbart LK, Sokolowsky S, Marienhagen J. Displaced by deceivers: prevention of biosensor cross-talk is pivotal for successful biosensor-based high-throughput screening campaigns. ACS Synth Biol 2019;8:1847–57. https://doi.org/10.1021/acssynbio.9b00149.

[91] Chen Y, Du L, Tian Y, Zhu P, Liu S, Liang D, et al. Progress in the development of detection strategies based on olfactory and gustatory biomimetic biosensors. Biosensors 2022;12:858. https://doi.org/10.3390/bios12100858.

[92] Cheng P, Xie X, Hu L, Zhou W, Mi B, Xiong Y, et al. Hypoxia endothelial cells-derived exosomes facilitate diabetic wound healing through improving endothelial cell function and promoting M2 macrophages polarization. Bioact Mater 2024;33:157–73. https://doi.org/10.1016/j.bioactmat.2023.10.020.

[93] Ausländer S, Ausländer D, Fussenegger M. Synthetic biology – the synthesis of biology. Angew Chem Int Ed 2017;56:6396–419. https://doi.org/10.1002/anie.201609229.

[94] Farzadfard F, Lu TK. Genomically encoded analog memory with precise in vivo DNA writing in living cell populations. Science 2014;346:1256272. https://doi.org/10.1126/science.1256272.

[95] Inniss MC, Silver PA. Building synthetic memory. Curr Biol 2013;23:R812–6. https://doi.org/10.1016/j.cub.2013.06.047.

[96] Gardner TS, Cantor CR, Collins JJ. Construction of a genetic toggle switch in *Escherichia coli*. Nature 2000;403:339–42. https://doi.org/10.1038/35002131.

[97] Elowitz MB, Leibler S. A synthetic oscillatory network of transcriptional regulators. Nature 2000;403:335–38. https://doi.org/10.1038/35002125.

[98] Ajo-Franklin CM, Drubin DA, Eskin JA, Gee EPS, Landgraf D, Phillips I, et al. Rational design of memory in eukaryotic cells. Genes Dev 2007;21:2271–76. https://doi.org/10.1101/gad.1586107.

[99] Ferrell JE. Self-perpetuating states in signal transduction: positive feedback, double-negative feedback and bistability. Curr Opin Cell Biol 2002;14:140–48. https://doi.org/10.1016/S0955-0674(02)00314-9.

[100] Boo A, Ledesma Amaro R, Stan G-B. Quorum sensing in synthetic biology: a review. Curr Opin Syst Biol 2021;28:100378. https://doi.org/10.1016/j.coisb.2021.100378.

[101] Miller MB, Bassler BL. Quorum sensing in bacteria. Annu Rev Microbiol 2001;55:165–99. https://doi.org/10.1146/annurev.micro.55.1.165.

[102] Kareb O, Aïder M. Quorum sensing circuits in the communicating mechanisms of bacteria and its implication in the biosynthesis of bacteriocins by lactic acid bacteria: a review. Probiotics & Antimicro Prot 2020;12:5–17. https://doi.org/10.1007/s12602-019-09555-4.

[103] Roy V, Adams BL, Bentley WE. Developing next generation antimicrobials by intercepting AI-2 mediated quorum sensing. Enzyme Microb Technol 2011;49:113–23. https://doi.org/10.1016/j.enzmictec.2011.06.001.

[104] Stephens K, Bentley WE. Synthetic biology for manipulating quorum sensing in microbial consortia. Trends Microbiol 2020;28:633–43. https://doi.org/10.1016/j.tim.2020.03.009.

[105] Didovyk A, Tonooka T, Tsimring L, Hasty J. Rapid and scalable preparation of bacterial lysates for cell-free gene expression. ACS Synth Biol 2017;6:2198–208. https://doi.org/10.1021/acssynbio.7b00253.

[106] Vendeville A, Winzer K, Heurlier K, Tang CM, Hardie KR. Making "sense" of metabolism: autoinducer-2, LUXS and pathogenic bacteria. Nat Rev Microbiol 2005;3:383–96. https://doi.org/10.1038/nrmicro1146.

[107] Pereira CS, Thompson JA, Xavier KB. AI-2-mediated signalling in bacteria. FEMS Microbiol Rev 2013;37:156–81. https://doi.org/10.1111/j.1574-6976.2012.00345.x.

[108] Alencar VC, Silva J, de F dos S, Vilas Boas RO, Farnézio VM, de Maria YNLF, Aciole Barbosa D, et al. The quorum sensing auto-inducer 2 (AI-2) stimulates nitrogen fixation and favors ethanol production over biomass accumulation in *Zymomonas mobilis*. Int J Mol Sci 2021;22:5628. https://doi.org/10.3390/ijms22115628.

[109] Wang B, Kitney RI, Joly N, Buck M. Engineering modular and orthogonal genetic logic gates for robust digital-like synthetic biology. Nat Commun 2011;2:508. https://doi.org/10.1038/ncomms1516.

[110] Singh V. Recent advances and opportunities in synthetic logic gates engineering in living cells. Syst Synth Biol 2014;8:271–82. https://doi.org/10.1007/s11693-014-9154-6.

[111] Chuang C-H, Lin C-L. Synthesizing genetic sequential logic circuit with clock pulse generator. BMC Syst Biol 2014;8:63. https://doi.org/10.1186/1752-0509-8-63.

[112] Bressler EM, Adams S, Liu R, Colson YL, Wong WW, Grinstaff MW. Boolean logic in synthetic biology and biomaterials: towards living materials in mammalian cell therapeutics. Clin Trans Med 2023;13:e1244. https://doi.org/10.1002/ctm2.1244.

[113] Buchler NE, Gerland U, Hwa T. On schemes of combinatorial transcription logic. Proc Natl Acad Sci 2003;100:5136–41. https://doi.org/10.1073/pnas.0930314100.

[114] Chen Y, Ho JML, Shis DL, Gupta C, Long J, Wagner DS, et al. Tuning the dynamic range of bacterial promoters regulated by ligand-inducible transcription factors. Nat Commun 2018;9:64. https://doi.org/10.1038/s41467-017-02473-5.

[115] Cubillos-Ruiz A, Guo T, Sokolovska A, Miller PF, Collins JJ, Lu TK, et al. Engineering living therapeutics with synthetic biology. Nat Rev Drug Discov 2021;20:941–60. https://doi.org/10.1038/s41573-021-00285-3.

[116] Wong A, Wang H, Poh CL, Kitney RI. Layering genetic circuits to build a single cell, bacterial half adder. BMC Biol 2015;13:40. https://doi.org/10.1186/s12915-015-0146-0.

[117] Taton A, Ma AT, Ota M, Golden SS, Golden JW. NOT gate genetic circuits to control gene expression in cyanobacteria. ACS Synth Biol 2017;6:2175–82. https://doi.org/10.1021/acssynbio.7b00203.

[118] Nielsen AA, Voigt CA. Multi-input CRISPR/Cas genetic circuits that interface host regulatory networks. Mol Syst Biol 2014;10:763. https://doi.org/10.15252/msb.20145735.

[119] Liu Y, Zeng Y, Liu L, Zhuang C, Fu X, Huang W, et al. Synthesizing AND gate genetic circuits based on CRISPR-Cas9 for identification of bladder cancer cells. Nat Commun 2014;5:5393. https://doi.org/10.1038/ncomms6393.

[120] Alberts B, Johnson A, Lewis J, Raff M, Roberts K, Walter P. Site-specific recombination. Molecular Biology of the Cell. 4th edition, Garland Science; 2002.

[121] Brenner's Encyclopedia of Genetics. 2013.

[122] Wang Y, Yau -Y-Y, Perkins-Balding D, Thomson JG. Recombinase technology: applications and possibilities. Plant Cell Rep 2011;30:267–85. https://doi.org/10.1007/s00299-010-0938-1.

[123] Weinberg BH, Pham NTH, Caraballo LD, Lozanoski T, Engel A, Bhatia S, et al. Large-scale design of robust genetic circuits with multiple inputs and outputs for mammalian cells. Nat Biotechnol 2017;35:453–62. https://doi.org/10.1038/nbt.3805.

[124] Schweizer HP. Applications of the *Saccharomyces cerevisiae* Flp-FRT system in bacterial genetics. J Mol Microbiol Biotechnol 2003;5:67–77. https://doi.org/10.1159/000069976.

[125] Anastassiadis K, Fu J, Patsch C, Hu S, Weidlich S, Duerschke K, et al. Dre recombinase, like Cre, is a highly efficient site-specific recombinase in *E. coli*, mammalian cells and mice. Dis Model Mech 2009;2:508–15. https://doi.org/10.1242/dmm.003087.

[126] Siuti P, Yazbek J, Lu TK. Synthetic circuits integrating logic and memory in living cells. Nat Biotechnol 2013;31:448–52. https://doi.org/10.1038/nbt.2510.

[127] Cubillos-Ruiz A, Guo T, Sokolovska A, Miller PF, Collins JJ, Lu TK, et al. Engineering living therapeutics with synthetic biology. Nat Rev Drug Discov 2021;20:941–60. https://doi.org/10.1038/s41573-021-00285-3.

[128] Yan X, Liu X, Zhao C, Chen G-Q. Applications of synthetic biology in medical and pharmaceutical fields. Sig Transduct Target Ther 2023;8:1–33. https://doi.org/10.1038/s41392-023-01440-5.

[129] Wurtzel ET, Vickers CE, Hanson AD, Millar AH, Cooper M, Voss-Fels KP, et al. Revolutionizing agriculture with synthetic biology. Nat Plants 2019;5:1207–10. https://doi.org/10.1038/s41477-019-0539-0.

[130] Sargent D, Conaty WC, Tissue DT, Sharwood RE. Synthetic biology and opportunities within agricultural crops. J Sustain Agric Envir 2022;1:89–107. https://doi.org/10.1002/sae2.12014.

[131] Yang J-S, Reyna-Llorens I. Plant synthetic biology: exploring the frontiers of sustainable agriculture and fundamental plant biology. J Exp Bot 2023;74:3787–90. https://doi.org/10.1093/jxb/erad220.

[132] Le Feuvre RA, Scrutton NS. A living foundry for synthetic biological materials: A synthetic biology roadmap to new advanced materials. Synth Syst Biotechnol 2018;3:105–12. https://doi.org/10.1016/j.synbio.2018.04.002.

[133] Scown CD, Keasling JD. Sustainable manufacturing with synthetic biology. Nat Biotechnol 2022;40:304–07. https://doi.org/10.1038/s41587-022-01248-8.

[134] Lobsiger N, Stark WJ. Strategies of immobilizing cells in whole-cell microbial biosensor devices targeted for analytical field applications. Anal Sci 2019;35:839–47. https://doi.org/10.2116/analsci. 19R004.

[135] Chong H, Ching CB. Development of colorimetric-based whole-cell biosensor for organophosphorus compounds by engineering transcription regulator DmpR. ACS Synth Biol 2016;5:1290–98. https://doi.org/10.1021/acssynbio.6b00061.

[136] Ho JML, Miller CA, Parks SE, Mattia JR, Bennett MR. A suppressor tRNA-mediated feedforward loop eliminates leaky gene expression in bacteria. Nucleic Acids Res 2021;49:e25. https://doi.org/10.1093/nar/gkaa1179.

[137] Wang B, Barahona M, Buck M. A modular cell-based biosensor using engineered genetic logic circuits to detect and integrate multiple environmental signals. Biosens Bioelectron 2013;40:368–76. https://doi.org/10.1016/j.bios.2012.08.011.

[138] Aoki SK, Lillacci G, Gupta A, Baumschlager A, Schweingruber D, Khammash M. A universal biomolecular integral feedback controller for robust perfect adaptation. Nature 2019;570:533–37. https://doi.org/10.1038/s41586-019-1321-1.

[139] Green AA, Kim J, Ma D, Silver PA, Collins JJ, Yin P. Complex cellular logic computation using ribocomputing devices. Nature 2017;548:117–21. https://doi.org/10.1038/nature23271.

[140] Zadeh JN, Steenberg CD, Bois JS, Wolfe BR, Pierce MB, Khan AR, et al. NUPACK: Analysis and design of nucleic acid systems. J Comput Chem 2011;32:170–73. https://doi.org/10.1002/jcc.21596.

[141] Kawasaki S, Ono H, Hirosawa M, Kuwabara T, Sumi S, Lee S, et al. Programmable mammalian translational modulators by CRISPR-associated proteins. Nat Commun 2023;14:2243. https://doi.org/10.1038/s41467-023-37540-7.

Merve Yavuz, Ahmet Hınçer, Aslı Semerci,
and Urartu Özgür Şafak Şeker

Chapter 2
Biological devices for cellular targeting and decision-making

Abstract: The engineered biological systems offer new avenues for precision medicine and therapeutic interventions. The development of biological devices for cellular targeting and decision-making executes synthetic biology tools to integrate computational logic and sensing capabilities in living cells. The machineries have the ability to detect specific biomarkers, process environmental signals, and perform targeted responses. In order to specifically recognize and destroy cancer cells, one of the crucial advancements resides in the design of chimeric antigen receptor T cells. The cells harbor synthetic genetic circuitries to sense external stimuli, the presence of a pathogen or a disease marker, and to give a cellular response. The biological devices are designed as robust and adaptive based on synthetic biology applications by leveraging modular and programmable genetic components. The efficacy and safety of cellular therapies are enhanced via precise cellular targeting and dynamic decision-making that provide promising approaches for diagnosing and treating a wide range of diseases and personalized medicine.

2.1 Biological devices in cellular targeting

Biological devices are engineered to produce precise targeting machineries against specific cells or cellular components. Innovative solutions for therapeutic developments can be offered via cutting-edge fields within synthetic biology and biotechnology. The overarching goal is to design cellular machineries for specific targeting of cells, allowing targeted therapeutic agent delivery and specific pathway modulation opportunities.

2.1.1 Significance of targeting specific cells

With the help of synthetic biology tools, customizable drug delivery vehicles can be designed, like engineered bacteria or viruses, for targeted therapeutic delivery. Pre-

Merve Yavuz, Ahmet Hınçer, Aslı Semerci, and Urartu Özgür Şafak Şeker, UNAM–Institute of
Materials Science and Nanotechnology, Bilkent University, Ankara 06800, Turkey

https://doi.org/10.1515/9783111329499-002

cise drug release can be achieved upon the response of programmed vehicles to a specific cue in the microenvironment.

The engineering of nanoparticles is utilized for specific drug delivery, even considering precision, immunogenicity, and toxicity. In contrast, utilizing live cells that inherently possess a targeting mechanism can serve as effective carriers for drug delivery. The accumulation of the drug in the desired area is enhanced. If the native cell type lacks the required therapeutic characteristics, the genetic circuit integration to reprogrammed cells improves the yield of desired outcomes. To advance drug delivery, various cell types, such as bacteria, yeast, and red blood cells, can be utilized whenever needed [1].

The construction of logic gates within therapeutic cells by synthetic biology tools provides the processing of input signals and stimulates specific therapeutic response production as an output if certain conditions are present. The strategy provides enhanced precision for cellular targeting applications.

2.1.2 Cellular targeting strategies and applications

2.1.2.1 Ligand–receptor interactions

Affibodies are small molecules engineered for binding to target proteins or peptides with high affinity and selectivity [2]. Bacterially derived outer membrane vesicles (OMVs) can be engineered to reduce endotoxicity toward human cells by introducing a mutation for their use in active tumor-targeting applications. By introducing a particular mutation to the *Escherichia coli* K-12 W3110 strain, lipopolysaccharide (LPS) modifications lead to reduced endotoxicity. Incorporation of a cancer-specific ligand, HER2-specific affibody, into OMVs enables targeted payload delivery in a cell-selective manner for targeted gene-silencing and tumor growth regression. This brings various advantages like reduced side effects due to specificity, low endotoxicity, the capacity to encapsulate siRNA, and the potential for large-scale production through cost-effective bacterial fermentation. The payload could be arranged according to the needs; chemotherapeutic agents can be loaded and subjected to in vivo testing [3].

Nanoparticles can be engineered to be combined with biological components like proteins or peptides by using synthetic biology tools in order to interact specifically with target cells and facilitate drug delivery. They have been extensively utilized for improving the effectiveness of drugs. However, there might be some insufficiencies of the nanoparticles, like that a small percentage of them can effectively reach the intended sites, cell–nanoparticle interactions can lead to toxic effects, achieving cell selectivity in heterogeneous diseased tissues like chronic wounds and cancer poses challenges, phagocytic clearance results in nanoparticle depletion, and there has been limited success in clinical applications [1].

Nanobiotechnology presents numerous opportunities for delivering synthetic biology devices to meet the need for targeted delivery, which is a necessity in the majority of in vivo theranostic approaches [4]. DNA aptamers can serve as activators for cargo release from delivery systems with a strong binding affinity for a specific protein. DNA origami "nanorobots" were engineered to transport payloads like gold nanoparticles or antibody fragments labeled with fluorescence to create the concept where the releasing molecules initiate cell signaling. The aptamer-encoded logic gate approach could inspire novel machineries featuring selectivity, and biologically active payloads serve for cell targeting applications [4, 5].

2.1.2.2 Cellular markers and recognition

Engineered immune cells with synthetic biology tools can be designed to enhance their targeting capabilities. This includes modifying natural killer cells or macrophages to improve their ability to recognize and eliminate specific target cells.

Chimeric antigen receptor (CAR)-T-cell (CAR-T) therapy aims to target specific cancer cell antigens by employing genetically modified T cells to express a synthetic receptor (CAR). The ability to specifically target and destroy cancer cells through immunity is improved via synthetic biology approaches. The modified T cells leverage the treatment landscape for hematologic malignancies. For CAR-T therapies to be considered for solid tumors, the occurrence of on-target, off-tumor toxicities must be addressed to eliminate the fatal side effects in healthy tissues. The toxicity can be mitigated by including immunosuppressive corticosteroid therapy and suicide-gene engineering. To improve specificity and reduce off-tumor toxicity by controlling CAR activation, sophisticated genetic circuits like inhibitory CARs or synthetic Notch (SynNotch) receptors are developed. The crucial step is identifying the optimal antigens for differentiating solid tumors from healthy tissues, especially in tissue injury. The need for precise control of CAR-T activation in solid tumors could be achieved via mechanosensory-engineered cells coupled with microbubbles that amplify ultrasound signals. In vivo applications may be hurdled due to microbubble requirements. An inducible CAR-T controlled via focused ultrasound (FUS) is developed to activate the engineered T cells at specific times and locations via short-pulsed FUS and effectively suppress tumor growth [6]. Similarly, regulation of CAR-T photothermally showed over 60-fold higher expression of a reporter transgene without affecting key functions of primary T cells. The designed synthetic gene switch activates transgene expression upon sensing a mild temperature increase (40–42 °C). The on-target activity, expression of the transgene specifically within tumors, is achieved via heating CAR-T photothermally with the help of gold nanorods. The localized production of an interleukin-15 superagonist or a bispecific T-cell engager targeting the natural killer group 2D ligand (NKG2DL) increases antitumor activity [7]. As an alternative to heat, a light-inducible nuclear translocation and dimerization (LINTAD) system has been developed to precisely control gene regulation and CAR-T activation. Potent cytotoxicity against cancer cells is developed for precision cancer immunotherapy via pulsed light stimulation activating LINTAD CAR-T [8].

In developing "safe and live drugs" for targeted cancer therapy, bacterial cells are engineered to create chromosome-free, lack of uncontrolled bacterial growth SimCells and mini-SimCells. A colorectal cancer biomarker is targeted via surface display of anti-carcinoembryonic antigen nanobody display on engineered cellular machineries. It is shown that the bacterial cells can specifically bind to the tumor cell and show cancer-killing effects via catechol synthesis with the genetic circuit, a potent anticancer compound. The proposed chromosome-free and nondividing cells are considered safe and could potentially be used in novel bacteria-derived cancer treatments (Figure 2.1) [9].

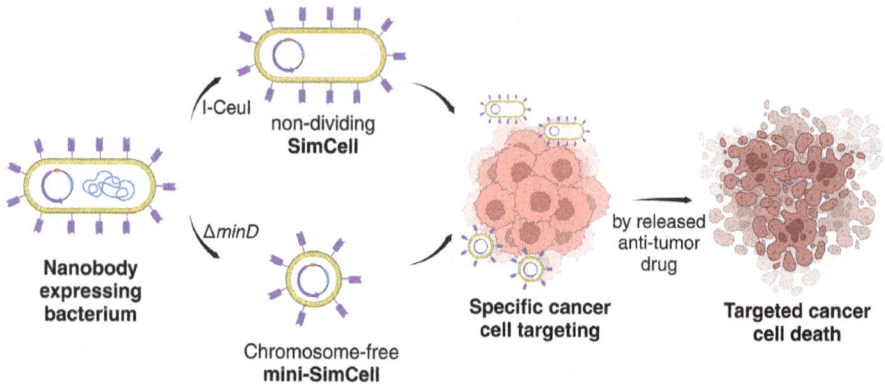

Figure 2.1: Schematic for representing reprogrammed bacteria for specific cancer targeting and therapeutic release via synthetic biology tools. The engineered chromosome-free, nondividing bacterial cells display nanobodies on the cell surface to bind the biomarker commonly found in colorectal cancer cells. Inducible bacterial cells have the ability to produce anticancer compounds that provide controllable delivery of therapeutics into the tumor microenvironment (reproduced from [9]; created with BioRender.com).

Phage-based diagnostic approaches offer distinct advantages over cell-based biosensors that generally meet the clinical application requirements. Phages are viruses that target bacteria through specific host cell surface receptors. In a straightforward model, reporter phages are designed so that the presence of the correct epitope prompts the transfer of the phage genome into bacteria, initiating bioluminescence via luciferase in the targeted bacteria. Many pathogen types can be diagnosed via bioluminescent reporter phage application including *Bacillus anthracis*, *Salmonella typhimurium*, and *Staphylococcus aureus* [10].

In addition to diagnostics, phages have potential applications that can be used as bioelectrode elements either by immobilizing them onto electrode surfaces or utilizing lytic phages to infect target bacteria and release bacterial cell contents like ATP for electrochemical detection. An electrode modified with M13 bacteriophage was developed to detect *Salmonella* spp. via binding to target bacteria through amino acid groups on the filamentous phage. Alternatively, lytic phages specific to *S. aureus* were integrated into a surface-modified bacterial cellulose matrix. This enables the detection *of S. aureus* via impedance measurements. The developed biosensors, consisting of bacterial cells and phages, have promising potential in efficiently detecting disease-associated biomarkers for medical diagnosis [10].

Detecting biological substances via whole-cell biosensors needs to resolve the low-signal-to-noise ratio to be used for medical diagnosis in clinical applications. The genetically encoded digital amplifying genetic switch utilized bacterial machineries are designed for biomarker detection in human urine and serum. The Boolean logic gate is utilized for computational processes of bactosensors. The multifunctional sensor can perform signal digitalization, multiplexed signal processing, signal amplification, and data storage. A specific biomarker can be detected at a predefined threshold value, and colorimetric output enables medical decisions. The bacteria sensor machinery enables the multiplexed detection of pathological biomarkers. The Boolean integrase logic gate module assures signal digitalization, user-defined medical algorithm, and storage of diagnostic test outcomes. The module can be reprogrammed for different medical detection purposes based on highly robust whole-cell biosensors tested to detect pathological glycosuria with clinical samples [11].

Bacteria-based microrobots were envisioned as antitumor agents by being designed for theranostic purposes of targeting and treating solid tumors. The attenuated *S. typhimurium* bacterial cells harbor a high affinity to bind Cy5.5-coated polystyrene microbeads via biotin–streptavidin interaction. The microstructure is a therapeutic molecule carrier that makes bacteriobots an active drug delivery system (DDS). The intrinsic propensity of the bacterial cell to chemotaxis via their chemotactic receptor and flagella, the solid tumors can be targeted. The programmed bacteriobots in a chemotactic microfluidic device responded to the concentration gradient of tumor lysates or spheroids by moving from the central region to the side regions at a higher velocity than normal cells. With the injection of bacteriobots into a CT-26 tumor mouse model, their chemotactic motility and tumor-targeting capability were ensured via Cy5.5 signal detection at the tumor site. Both in vitro and in vivo tests showed an innovative approach to using bacteria as a microactuator and microsensor for carrying microstructures to tumors [12].

The diagnosis of a specific disease marker can be achieved via engineering the bacteria colonizing specific tissue which enables targeted therapeutic interventions. The probiotic-based approach can be leveraged to enhance colonization resistance and serve as an early-stage diagnostic tool for cholera. As a microbiota-modulating intervention, oral administration of a common fermentative bacterium, *Lactococcus lactis*, reduced the *Vibrio cholerae* pathogen colonization and improved survival rates in infected infant mice. The intrinsic property of *L. lactis* is to harbor high carbon flux metabolism and give strong acidification via lactic acid production, which plays an antagonistic agent role for *V. cholerae*. In addition, by designing synthetic gene circuits, the lactic acid bacteria were engineered to detect the quorum-sensing signals from *V. cholerae* in the gut and stimulate enzymatic reporter expression identified in fecal samples. The dietary involvement of both natural and engineered *L. lactis* can impede cholera progression and improve disease surveillance in populations at risk of cholera outbreaks [13].

2.1.2.3 Tissue-specific promoters

The engineering of genetic circuitries with custom-designed promoters enables the specific signal or condition detection and provides a response. The cells are designed to harbor synthetic components to control the expression of therapeutic genes that execute targeted responses. Microorganisms might have an inclination to migrate and accumulate in tumor cells. With this ability, they can be modified to evaluate and report the status of tumors. The colonization of the microorganism can be facilitated by low oxygen levels, an irregular blood supply or terminate nutrient availability. In a particular investigation, nonpathogenic commensal bacteria underwent genetic modifications to express the luxCDABE operon, providing bioluminescent capabilities. The combination of these strains with bioluminescent 3D optical tomography and micro-CT scanning allows for the visualization of the malignant growths when introduced into mice with subcutaneous xenograft tumors. By utilizing heme-binding proteins and other reporters derived from functional magnetic resonance imaging, the expression of protein-based contrast agents could potentially relay information from deeper tissues, utilizing orthologous reporters for enhanced data retrieval [4].

A probiotic bacterium, PROP-Z, was engineered to express lacZ and the luxCDABE operon. These modified bacteria were employed to precisely target, visualize, and diagnose liver metastasis. A dual maintenance strategy, including the expression of the *Bacillus subtilis* alp7 gene (resulting in filament production for equal plasmid segregation during cell division), as well as a toxin–antitoxin system compelling the cell to either maintain the plasmid or face cell death, is used to ensure the stability of the synthetic circuitry within PROP-Z. Bioluminescent imaging can be done by administering the bacteria orally or intravenously that exhibit a preference for colonizing liver or subcutaneous tumors [4].

An engineered therapy harnessing bacteria to release cytotoxic agents on-site is designed based on the bacterial synchronous lysis cycle at a specific population density. Both the needs for drug loading and releasing mechanisms are eliminated with this engineered system via expressing the pore-forming antitumor toxin under a quorum lysis promoter. The coculturing of bacteria with human cancer cell lines on a microfluidic device showcases the potential usage of the system over conventional cancer therapies. Combinatorial therapy of circuit-engineered bacteria and chemotherapy reduces the tumor activity when the strain is orally administered to a syngeneic mouse transplantation model of hepatic colorectal metastases [14].

Optogenetic tools, which harbor light-sensitive proteins, can be engineered into cells to control cellular processes with light. This synthetic biology approach allows for precise spatiotemporal control over therapeutic interventions. The garnered attention toward bacteria-based cancer therapy stands due to bacteria's specific tumor-targeting ability and antitumor immunity activation. Besides that, tyrosinase has the intricate property to use cyanine 5 tyramides (Cy5-Tyr) as a substrate and synthesize Cy5-labeled melanin-like polymers (Cy5-Mel). This noteworthy process is an antitumor activity when tyrosinase is overexpressed in a probiotic bacterium, *E. coli* Nissle (EcN) 1917. The resultant bacterial machineries show strong fluorescence administered for guidance and significant photothermal properties when near-infrared (NIR) laser irradiated. The tumor-colonizing bacteria act as an adjuvant that triggers immunogenic cell death of tumor cells and dendritic cells (DCs) maturation upon laser exposure on the tumor site [15]. Using bacteria as an adjuvant may also serve as an opportunity to overcome the hurdle of the immunosuppressive nature of the tumor microenvironment in a way that activates the immune response of tumor antigens released upon radiotherapy or chemotherapy. The highly motile *Salmonella* VNP20009 attenuated strain is engineered for dual activity where it can adsorb the released tumor antigens following their transport out of the tumor core and can enhance the immunogenicity of antigens via its binding capability to toll-like receptors (TLRs). The bacterial strain coated with cationic polymer nanoparticles for antigen absorption enables the stimulation of marginal DCs in the tumor periphery harnessed as an antitumor agent having potential usage for in situ cancer vaccination (Figure 2.2) [16].

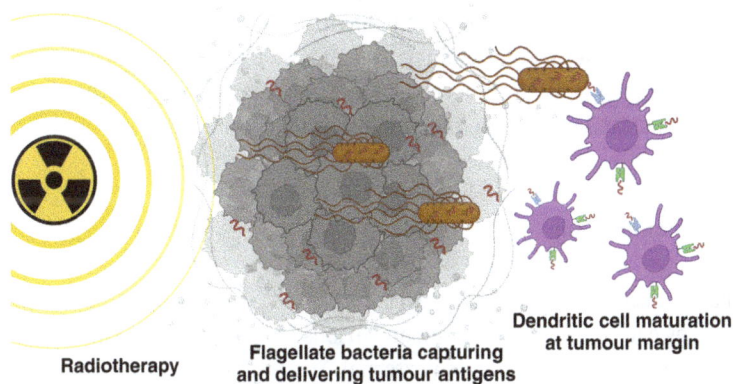

Figure 2.2: The representation of antigen-capturing bacteria upon radiation therapy for cancer.
The intratumorally injected-flagellate bacteria are engineered to adsorb tumor antigens and transport them to antigen-presenting cells at the tumor margin (reproduced from [16]; created with BioRender. com).

The oral DNA vaccination harboring attenuated bacteria is implicated in overcoming low infection efficiency in cancer immunotherapy. The coating of *Salmonella* bacteria with a synthetic nanoparticle self-build from cationic polymers and plasmid strategy is adopted to create a protective layer against the acidic environment of the stomach and intestines. In addition, the engineered bacterial vectors that encode autologous vascular endothelial growth factor receptor 2 induce T-cell activation and cytokine production, inhibiting tumor growth [17].

2.2 Decision-making in cellular systems

2.2.1 Cellular decision-making and information processing

In nature, cellular decisions are often made stochastically, meaning they are subject to random variations and probabilities. Despite this inherent randomness, these decisions typically operate in a binary manner. The cell processes information from its environment and internal signaling pathways in a way that can be likened to threshold testing. When a signal reaches a certain critical level, it triggers a binary decision within the cell. Consequently, the cell adopts the necessary characteristics, which are then manifested both genotypically and phenotypically. This binary decision-making process is crucial for various cellular functions and behaviors. For example, a cell might decide whether to grow, divide, or initiate programmed cell death in response to environmental stimuli such as nutrient availability or stress conditions. These decisions are fundamental to the survival and adaptation of organisms [18, 19].

Understanding the stochastic nature and binary decision-making of cellular processes provides valuable insights into cellular behavior and can inform various research fields, including developmental biology, cancer research, biotechnology, and synthetic and systems biology. By deciphering these mechanisms, targeted therapies and innovative solutions can be developed for complex biological problems.

2.2.1.1 Basis of decision-making

To explain the basic mechanism of cellular decision-making, the different orders of decisions a cell makes can be examined based on various cues. These cues can be categorized into two main sources: internal and external. The integration of these signals enables the cell to comprehend its current state and anticipate future conditions, thereby increasing the likelihood of making appropriate decisions for both present and future states [20].

Internal cues, often creating cellular noise, can be likened to a ball perched at the top of a hill (Figure 2.3). In this analogy, cells are balanced dynamically to alter their features with even a slight nudge from a signal. This readiness contributes to the stochastic nature of decision-making processes to a certain extent. The cell's response to signals, and the consequent momentary fluctuations in the levels of cellular components, may lead to incorrect responses in the short term. However, there is a consistent

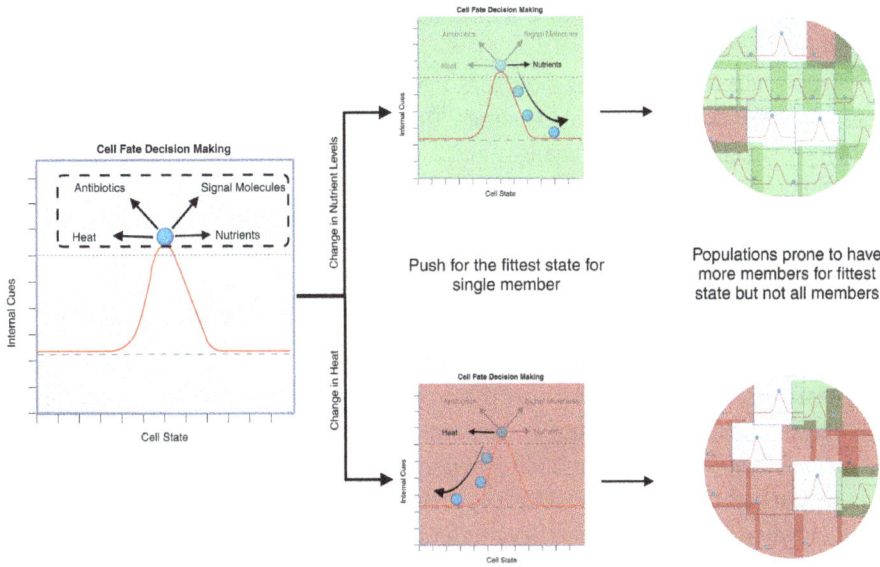

Figure 2.3: Cell fate decision-making analogy. A cell is constantly producing internal cues to be responsive even for a small change in environment. One "push" coming from the environment can lead to a change in cell state to the fittest version; however, the stochasticity in the cell itself or environment may create exceptional cell states. In the population aspect, the probability of having the members with the fittest state is enriched, although other states are still found. In the end, it seems like a binary decision rather than pure randomization (created with Biorender.com).

directional push, such as the presence of a sugar source in the environment, which can make the population decisively act in harmony. Among all possible states, the cell adopts the fittest state, leading to a binary decision-making process rather than pure randomness [20, 21].

This "ball at the top of the hill" analogy also effectively illustrates the persistence of decisions across future generations. Once the initial push occurs, a series of sequential events result in a population with more stabilized features. This population inherits a predisposition toward the initial decision, effectively transmitting information. While adapting to environmental factors is crucial, this does not eliminate the cell's ability to make new decisions entirely. When signals change, the population is ready to ascend the hill again in search of new stimuli.

This dynamic is not deterministic; it is a simultaneous process. For instance, when sugar becomes abundant in the environment, the population's characteristics shift to optimize sugar utilization. Over time, as the sugar gets depleted, the cells become primed again, ready for the next signal. A similar principle applies to antibiotic exposure. Although the population responds to environmental change – in this case, the presence of antibiotics – they retain their adaptability. If the antibiotic pressure is removed, they begin to repopulate with traits similar to those prior to the antibiotic exposure [22].

Cellular decision-making is a complex interplay of internal and external cues, stochastic processes, and adaptive responses. This system ensures that cells can respond to immediate environmental changes and maintain the flexibility to adapt to future conditions.

2.2.1.2 Cell fate decision-making applications

Both prokaryotic and eukaryotic cells possess intrinsic mechanisms for making cell-fate decisions. A notable example of this in prokaryotic cells is the quorum-sensing mechanism. Environmental cues or decisions made by other individuals in the population can alter the behavior of both individual cells and groups. A prime example of this system can be observed in *Vibrio* spp. Depending on the population density and resulting changes in autoinducer levels, these bacteria adjust their behavior to function either as individuals or as a collective group. In high-density populations, the increased concentration of autoinducers affects receptors such as LuxN, LuxPQ, and CqsS. When these receptors interact with autoinducers, they become phosphatases, removing phosphate groups from signal transduction elements like LuxO and LuxU. Normally, these elements are phosphorylated and promote the transcription of AphA, which mediates individual behavior. The dephosphorylation of these elements leads to the transcription of LuxR or HapR, which mediate group behavior. This process is analogue to harnessing the "wisdom of the crowds" [23].

In *Pseudomonas putida*, quorum-sensing induces population heterogeneity. When autoinducer levels do not reach a critical threshold, group behavior does not manifest, resulting in a bistable response that protects against sudden environmental changes. This adaptive mechanism allows the population to alter its decisions based on current needs [24]. Another study demonstrated that quorum sensing helps cells understand environmental cell density and diffusibility, optimizing the production rates of costly molecules [25].

Controlling cell fate artificially has also been explored. For instance, adding a synthetic negative feedback loop circuit to *E. coli* and altering its conditions can regulate population density. This system exploits the natural LuxI/LuxR quorum sensing mechanism of *Vibrio fischeri* to function in *E. coli* populations. The system facilitates the production of acyl-homoserine lactone (AHL) signaling molecules. As AHL molecules accumulate due to increasing cell density, they activate the synthetic circuit, producing a killer protein dependent on AHL levels. This precise control mechanism alters the population density and fate of the cells, independent of environmental sources. Such control and precision with a synthetic circuit hold potential for microbiome engineering as a therapeutic strategy [26].

In eukaryotic systems, especially within multicellular organisms, cellular decision-making can have profound and immediate effects on the fate of the entire organism. A single decision made by a cell can influence subsequent developmental pathways, fundamentally altering the future branches of the possible fate tree.

For example, during embryonic development, changes in morphogenetic cues can significantly alter the cell types into which progenitor cells will differentiate. This process highlights the sensitivity and complexity of developmental cues in determining cell fate [27]. However, advancements in induced pluripotent stem cell technology have demonstrated that it is possible to revert differentiated cells back to a pluripotent state, effectively resetting their developmental potential to a certain extent. This technology showcases the fluidity of cell fate and the potential for reprogramming cells [28].

Neural plasticity provides another example of dynamic decision-making within eukaryotic systems. The ability of neural tissue to adapt and reorganize in response to environmental cues allows for regaining lost functions through the activity of different cell types. This plasticity is crucial for repairing and maintaining nervous system function and highlights the adaptability of biological systems in response to changing needs [29].

In single-celled eukaryotes like yeast, decision-making processes can also be controlled by constructing heat-responsive or inducer-responsive elements. These synthetic elements allow researchers to manipulate gene networks and cellular decisions in response to specific stimuli. Such engineered systems enable precise control over cellular behavior and demonstrate the potential for synthetic biology in understanding and directing cellular processes [30, 31].

In summary, the study of decision-making in prokaryotic and eukaryotic systems, from single cell to multicellular organisms, reveals the intricate interplay of environmental cues, genetic networks, and cellular plasticity. These insights both deepen our understanding of cellular behavior and also open up new avenues for therapeutic approaches and biotechnological applications.

2.2.2 Programmable functionalities

Cellular decision-making and information processing occur naturally within biological systems. Understanding how and why these processes happen is crucial, as it can lead to the ability to rewire signal transduction pathways to achieve specific functionalities for various tasks. This is where synthetic and systems biology approaches become invaluable. By hijacking natural systems for repurposing or introducing novel responsive elements to organisms that do not naturally possess them, cellular functions can be programmed effectively similar to programming computers.

This approach has led to significant advancements in various fields, such as biosensing for diagnostics, understanding disease mechanisms, drug-target identification, therapeutic treatments, and DDSs [32]. For instance, in the study of agammaglobulin-emia, a condition characterized by the lack of mature B cell production in humans, the human B cell receptor signaling pathway has been reconstructed to gain insights into the disease mechanism. In another study, by programming bacteria in mice using bacteriophages, the effects of antibiotic treatment have been enhanced to combat with antibiotic-induced resistance [33]. Moreover, the bacteria with a genetic circuit activated via environmental signals coming from tumor microenvironment and pro-ducing an invasive protein has been demonstrated to carry drugs to cancer cells spe-cifically [34].

Additionally, synthetic biology tools like RNA-based and gates have been utilized to reprogram tumor cells. These engineered cells can attract immune cells only when the specifically required signals are present, ensuring accurate targeting of the tumor cells. This precise control over cellular behavior demonstrates the potential of syn-thetic biology in developing innovative therapeutic strategies [35].

By leveraging these advanced techniques, cells with enhanced or entirely new functionalities can be designed. For example, biosensors developed through synthetic biology can detect specific biomarkers for early disease diagnosis, improving patient outcomes. Understanding disease mechanisms at a cellular level allows for identifying novel drug targets, facilitating the development of more effective treatments. More-over, synthetic biology enables the creation of therapeutic cells capable of delivering drugs directly to diseased tissues, minimizing side effects and improving treatment efficacy.

The natural processes of cellular decision-making and information processing are pivotal areas of study. These processes can be harnessed and reprogrammed through synthetic and systems biology to achieve desired outcomes, leading to groundbreaking advancements in medicine, biotechnology, and beyond. By continuing to explore and manipulate these mechanisms, new possibilities will be unlocked for enhancing human health and addressing complex biological challenges.

2.2.2.1 Advanced bioproduction of therapeutics

The capability to control cellular functions is revolutionizing therapeutic approaches. With precise control over the mechanisms of action, it is possible to create workflows optimized for minimal requirements and maximum efficiency. This perspective is es-sential for advancing bioproduction processes. One of the primary challenges in cell therapeutics is the lack of reproducibility. Overcoming the randomness caused by ge-netic noise requires strict control over when and how cellular systems are activated to achieve consistent results.

Various bottlenecks must be addressed in the bioproduction of therapeutics to ad-vance the field. Focusing on cell-based therapeutics, several key challenges emerge:

expanding the cells, genetically modifying them, and purifying the cells of interest to meet good manufacturing practices standards.

CAR-T therapy is a promising cell therapeutics. This therapy has seen significant advancements by implementing new techniques in the bioproduction process. These include using bioreactors for cell expansion, flow-through electroporation for gene modification, and label-free sorting for cell purification. Such innovations are crucial for standardizing the production process and reducing batch-to-batch variation. Bioreactors, for instance, provide a controlled environment for the large-scale expansion of CAR-T, ensuring consistent growth conditions and cell quality. Flow-through electroporation enhances the efficiency and precision of gene modification, allowing for more reliable genetic engineering of the cells. Label-free sorting techniques improve the purification process by accurately isolating the desired cells without the need for additional labeling steps, thereby maintaining cell integrity and function [36].

These advancements enhance the reproducibility and efficacy of cell-based therapies and contribute to the scalability and cost-effectiveness of production. Standardizing these processes makes it possible to produce high-quality therapeutic cells at a larger scale, making treatments more accessible to patients.

Controlling cellular functions with precision is transforming therapeutic approaches, particularly in the field of cell-based therapies like CAR-T therapy. Addressing the challenges in bioproduction through innovative techniques and technologies is essential for advancing the field, ensuring reproducibility, and optimizing workflows. These efforts are paving the way for more effective and accessible treatments, highlighting the potential of synthetic and systems biology in modern medicine.

2.3 Synthetic biology tools

2.3.1 Usage of electronic circuits as an analogy for decision-making

The operation of a living organism is based on various controlling mechanisms, which have been described earlier. One of the most valuable explanations was based on the information provided by life scientists on electrical circuits; today, we are still using this analogy to understand and control genetic information from transcription control of DNA to posttranslation control of secreted proteins. The Lac operon system was described by Jacob and Monod [37] more than 60 years ago. This breakthrough approach was validated with experimental conditions, and a new era was born for genetic information's controllability. After that, in 1973, Rene Thomas described the genetic terms with a Boolean logic system; as a result of the controller's situation, true or false answers are shown as responses of the system [38]. Controlled protein production performed by multiple controllers affects protein production; Boolean formalization has been used for designing conditional protein production genetic circuits [38, 39]. Synthetic biologists have been using these rules to produce protein inside cells and in cell-free conditions. The gene expression controlling systems provide programming of protein regulation. Synthetic and natural gene regulation systems can be used for different purposes. Programmable cell-based and cell-free biosensors offer various applications that can be adapted to commercial products.

Transcription control starts with controlling promoter and terminator regions; these parts of the genes provide an on/off signal to the start of gene expression and the strength of transcription, which have been providing controllability to the natural and synthetic genetic circuits [40, 41]. Promoter and terminator sequences have been shown to have differences in eukaryotic and procaryotic organisms, and these differences have significant effects on the transcriptional and posttranslational processes. The action showing parts of these nucleic acid or peptide molecules can be engineered to show the desired activity, which starts with their genetic information. A synthetic genetic circuit contains an organism-specific-engineered promoter region to control protein production [42, 43]. Overall, the modular structure explained in the prior studies is used to develop living therapeutics or biosensors produced by engineered organisms. The electrical circuit analogy in synthetic biology is one of the touchstones of this area, which is used for controlling all of the parameters as a component in the circuit, and protein production is provided by managing all of the components to provide a final product. An engineered organism can regulate the biological system with the programmed triggering mechanism to produce fated biological final products.

2.3.2 Controllable cellular systems: different stages of translation control

Cellular decision-making switches are selected from recognizable and affinity-showing parts of nucleic acids and proteins induced by different factors such as light, electricity, and heat. These parts can be modified according to the requirements of the system; thereof, the engineered organism starts to show a targeted aim. This controllability can be provided by modifying transcription factors, promoter regions, mRNA sequences, or posttranslational stages.

Transcription factor complexes are described as the main transcription starters in the mRNA-producing regions in the genome because of the necessity of RNA polymerase affinities to the promoter regions at the correct conformation. These transcription initiator sequences show differences in eukaryotes and prokaryotes; also, different cell types can show different transcriptional factors to specific marker's production, which can be evaluated as tissue differentiation pathways in the eukaryotic cells. These genetic sequences bind directly or indirectly to transcription factors in harmony to create a transcription factor complex near the methionine amino acid's codon on the transcription site [44–46].

Promoters are initiative gene sequences that provide transcription factors binding activity to start mRNA production with various transcription factors. These sequences were discovered on the simian virus 40 (SV40) genome, and the following findings show that these sequences can be 100–1,000 bp long, including the transcription factor binding sequences, founded on the upstream of the related gene, constitutive or inducible with different factors [47]. A eukaryotic promoter region contains a TATA box to define the transcription start site and a CAAT box, a GC box, and a CAP site to describe the RNA polymerase binding site [45, 48]. A prokaryotic promoter contains consensus sequences named the Pribnow box at −10 and −35 positions to initiate transcription [46, 49]. In addition to these, promoter regions off signals are controlled by DNA methylation (CpG islands) in both types of organisms [50, 51]. Some promoters have basal transcription states, which are described as leakiness and can cause uncontrolled gene transcription. This low-transcription level can affect biochemical reactions and stochastic approaches and this should be considered during experimental design [52]. Also, different promoters have their limits for transcription, which can affect their final product for continuing RNA production [53]. According to the regulatory architecture of promoter region, transcription factors binding time can change, and gene transcription can affect the response time of the gene [54]. The promoter selection is crucial for developing a gene circuit and making decisions. To describe the types of promoters, we will investigate this subject under two headlines (constitutive and inducible) and explain the subtypes. The constitutive promoters do not create differentiation to the gene expression with any regulation factor; because of that, constitutive promoters (and their engineered versions) are used for high-protein production requiring purposes against conditional inducible promoters. A well-designed inducible system can sense different inducing agents like light, sound, electricity, various types of chemicals, proteins, oligonucleotides, or small molecules (Figure 2.4).

The posttranscriptional strength of the produced mRNA needs to be considered to regulate the amount of the final product yield; the signaling sequences at the end of the mRNA also determine the ribosome binding affinity and mRNA stability. The secondary structure of mRNA determines the rearrangement of mRNA (intron splicing), degradation capacity, an association of mRNA binding proteins, and translational level [55].

In addition to these nucleic acid-based controlling mechanisms inside the cells, posttranslational modifications are another important regulation mechanism. Cellular proteins can be modified after translation to engage and disassociate processes according to specific signaling parts of the peptides. These specific signal sequences can be added to the producing proteins during the genetic circuit design to produce the target protein after the inducement of the inducing factor [56, 57].

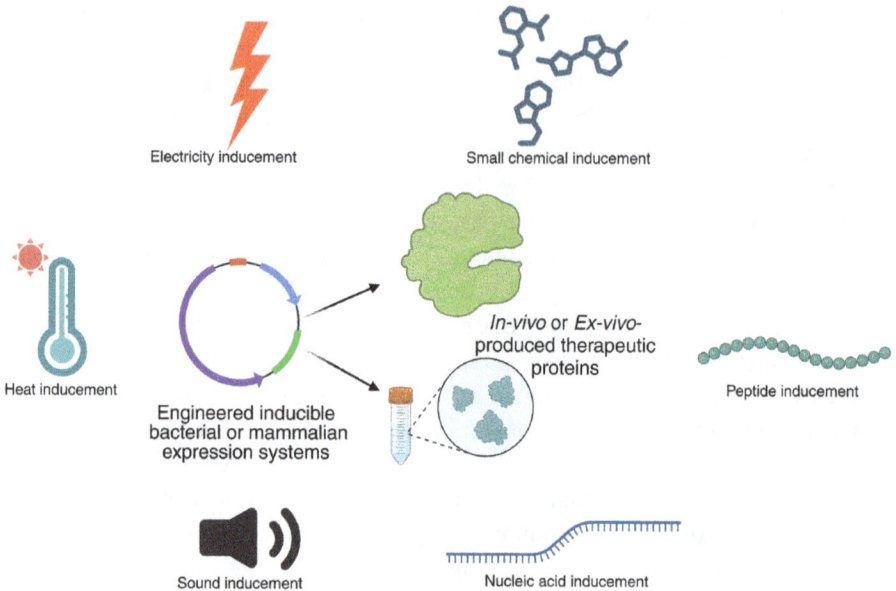

Figure 2.4: Different inducement factors of a programmed gene circuit. A programmed gene circuit can work inside the mammalian cell as a direct regulation system or a carrier organism. These factors can be triggered to induce target protein production, which is acting on the targeted region. Additionally, ex vivo protein production can be regulated with inducers to provide controlled protein production in a test tube or bioreactor (created with Biorender.com).

2.3.2.1 Inducible systems

Promoter activation is mainly related to transcription factors and enzymatic binding to the binding sites; any conformational unfavorability can cause stopping of gene expression [58]. Inducible promoters are found naturally in nature to provide cellular/organism-based homeostasis and adaptation to changeable environmental conditions [59]. The promoter activity can be controlled as a positive loop (i.e., LuxR) or negative loop (i.e., hormones) according to the type of promoters [58, 60]. Synthetic biologists develop targeted modifications on these natural promoter regions at the desired organism and are responsible for creating the desired circuits. With this strategy, modified organisms are controlled with inducible switches, which allow conditional gene transcription under specific situations like specific wavelengths of light, temperature, chemical inducers, etc. These living systems can be used for the targeted aims because of their responsive designed natures against changed homeostasis in a biological system. All of the determined inducers can act as starting signals on the gene transcription, which is useful for decision-creating systems starter inside the cell. The responsiveness advantage of these systems allows the production of biological materials and small molecules that can be used safely and with robust activity in different areas. In this part of the chapter, we will argue about various strategies for inducible systems are used for their clinical value in transcription and posttranslational levels.

2.3.2.1.1 Light-inducible systems

Light is a significant factor for maintaining cellular functions, especially polarization on membrane and protein affinity changes. Different wavelengths of light provide a wide range of physical rearrangements that control cell transcription, especially in light-sensitive organisms like plants [61]. Plants are naturally light-influenced systems with their photoreceptor activity, and these systems' considerable part of transcriptional activity is directly regulated by light. Different wavelengths of light can regulate the activity of plant genes in specific conditions; this genetic regulation activity can be applied to regulate other biological systems to create controllable systems [62]. The photoactivation-based genetic switches provide accurate control and sensitive characterizations of the designed systems through ion channel modification and recombinase enzyme activity control [63].

Optogenetics is one of the most critical developments in light-induced regulation of gene transcription. Optogenetics allows the activation of specific receptors, and deep brain activity investigations have provided control with optogenetic switches. Optogenetics describes photosensitive protein production used to control changes in the genetic circuits. This field opened the gates of brain mapping accomplishments by using different tools that have been used with synthetic biology products [64]. In the following years, light-inducible genetic circuits provided investigation opportunities for cell- and pathway-based investigations to understand many underlying mechanisms. The advantages of easily adapted and validated results of these tools have led to the development of new processes for applicational studies [65].

One of the crucial applications of optogenetic strategies is dynamically controlling microbial production and increasing the number of metabolites in industrial applications. Proteins, DNA, or RNA sequences can be targeted according to the systems' requirements. Photoreceptor protein's activation ability can provide an inducer effect according to the wavelength of the light, similar to the strategy used on plant and bacterial photoreceptors. These natural systems can be adapted to the desired organisms to take benefits of these biological processes. Using photoactive dimerized proteins can provide a production advantage in different ways, such as dimerization of proteins, gene activation starting, or transcription factor binding dynamics change. Many studies have focused on this controlling mechanism and taking advantage of cellular regulation control. The usage of *Arabidopsis thaliana*-originated phytochrome B (PhyB) and phytochrome-integrated factor 3 (PIF3) is one of the widely used control systems that can provide this aim in industrial and investigational applications. Raghavan and his colleagues have described the usage of an optogenetics circuit controlling in *E. coli* with the heterologous system controlling 660 nm light application to the cultured cells. This study produced an active T7 RNA polymerase as a two-part protein that controlled dimerization under PhyB and PIF3 photoactive components, which started the merging process after light emission [66]. The bacterial phytochrome biliverdin IXa is described as a bacterial NIR absorber and 740–720 nm NIR light activates this phytochrome. Kaberniuk et al.reported mammalian cells' optogenetic regulation with Idomarina species' photosensory core module usage, and the power of the experimental system was validated in cultured mammalian cells, primary isolated neurons, and intact tissues in mice. Results show that the usage of the photosensory core module provides significant activation time and protein yield changes in mammalian cells [67]. In both the systems, photoreceptor activation provides the assembly of subdomains of the protein and leads to the appearance of the final protein product at the end of the photoactivation process.

Using photoswitch systems in mammalian cells is another application of these controllable systems. A red light-responsive – *A. thaliana*-originated – phytochrome A promoter was recently adapted to mammalian cells to optically control Ras/Erk MAPK cascade by regulating the blood glucose level in mice, which can help control signaling pathways and epigenome regulation [68]. As a probiotic strain, EcN *1917* is also used for therapeutical purposes because of its toxicity-free cellular structure and modifiable genome. An engineered flaB promoter was developed to induce NIR inducing, which combined with TLR 5 to regulate innate response in the organism to cancer environment. The natural structural advantage of EcN strain is also providing a tumor migration capacity to this treatment strategy, and the study showed promising results of this system for usage in therapeutical applications [69]. These directly regulatable genetic circuits are prepared for use in daily life by different researchers; therefore, applications can be easily adapted to our lives in the recent feature with our smart-wearable devices such as smartwatches. An engineered EcN strain was combined with an optically controlling module to activate interleukin-10 secretion to treat ulcerative colitis. This approach provides real-time disorder monitoring with a highly robust reaction mechanism that can be adapted to home-care systems because of its applicational advantages [70]. Mansouri et al. developed a gene switch controlled by green light in the wavelength of smartwatches to regulate human glucagon-like peptide-1. Their green-light-activated synthetic promoters start the transcription of the locally engineered cells to treat diabetes in living organisms effectively [71]. A similar strategy was also used for developing smartphone flashlight-mediated triggering of the secretion of therapeutical molecules by the same researchers, and a repeatable and reversible secretion system was proposed [72].

The clustered regularly interspaced short palindromic repeats (CRISPR)-based photoactivation studies have also recently been described in the literature with these purposes; the robust-controlled on-off strategy of this method provides many advantages for creating rapid and reversible targeting of the gene of interest. To provide genomic regulation of the endogenous genome, targeted gene's promoter sequences are targeted with specific single-guide-RNAs, and transcription regulation can be controlled by the transcription factor's modification. This system's adaptation to the diagnostic tools has offered promising results with decreasing costs, shortened time, and robust detection.

Nihongaki et al. developed a photoreactivation system with this method, which can be counted as one of the leading studies of this era. This study used a catalytically inactivated dCas9 with light-sensitive cryptochrome 2 and combined with its binding partner calcium and integrin-binding protein 1 originated from *A. thaliana*. System activation allows the starting of gene transcription. The developed system provides an offer for adaptation to different promoter sequences by targeting the promoter with sgRNA [73]. After this study, the same strategy was applied to many other organisms and Cas proteins. In 2021, Zhou et al. [74] modified the pdDronpa domain because of the generalizable structure in different organisms. This improvement allows the production of single-strand Cas9 protein, which can be controlled with specific wavelength excitement. Different Cas enzymes were also used for light-targeted activation; Hu et al. proposed a platform with light-activated gRNA to diagnose viral infections. This system was proposed as a stable platform-like PCR system because of the use of Cas12 and Cas13 enzymes and the robust results [75]. Many studies suggest using CRISPR systems in diagnostic and therapeutic tools in the future; for example, using this system to control hunting in protein trafficking in the neuronal cells can provide solutions to this middle-age disorder forever [76]. Additionally, using light-activated CRISPR-based methods for neuronal reprogramming can be an efficient strategy for regenerative medicine. Shao et al. have developed a far-red light-controlled CRISPR-Cas9 activator system to control cell differentiation with gene expression control, which can be helpful in biomedical applications in the future [77].

Cre recombinase is one of the most studied photo-inducible components for synthetic biology approaches, and it has provided the development of essential mechanisms with protein engineering, photocaging, light-responsive carrier development, and light-induced expression control. These genetic regulation systems, especially highly specified cell lines like CAR-T, can provide beneficial results. For this purpose, a group of researchers developed CAR-T with tamoxifen-gated light-activated genetic circuits to control the T-cell with a Cre recombinase system with a combination of the binding affinity showing molecules of solid tumor cells. This photocaging system showed a high on-target off-tumor toxicity and enhanced therapeutic activity [78].

2.3.2.1.2 Oligonucleotide- and peptide-inducible systems

Regulation with nucleic acids and peptides is another critical and widely used mechanism for transcription control. Nucleic acid-based therapeutics have been used as approved drugs in the clinic since 1998. Using this technology with different approaches can provide advantages to developing next-generation therapeutics with their highly controllable mechanisms. Antisense nucleotides are one of the crucial tools to provide application advantages of these systems, which are based on targeting mRNA sequences for inhibiting translation, providing auspicious results in their clinical usage for monogenic disorders. Implementing electrical circuit analogy to nucleic acid-based methods is relatively easy because of their highly described structures. Recently,

Zhong et al. reported an engineered ribozyme as a reversible RNA switch to blocking mRNA expression in mammalian cells on anemia of chronic kidney disease. This promoter-independent system allows tissue-specific induction of various therapeutics in the same patient and has application advantages in cell-based gene therapies [79].

Another strategy for targeting a specific cell is targeting multiple markers simultaneously with protein-based Boolean logic systems including a protein marker and an mRNA. A recently published cancer drug development study applied the Boolean logic systems to peptide-based regulatable agents to create a stress call to innate immune metabolism. Three conformationally changeable peptides were designed to generate a complete structure to start signaling when they bind to the surface of the cancer cells. Also, an internal controlling mechanism was included in the design, which proceeded with particular competition to bind the targeted marker, and this cumulative activity provides preventing of off-targets in the mixed cell populations [80].

2.3.2.1.3 Chemically inducible systems

Small molecule-dependent gene regulation offers a wide range of opportunities to regulate gene expression of a gene with different targetable parts of transcription. Regulating promoters and inhibiting or activating transcription factors with small molecules can be counted for genetic manipulation methods with direct transcriptional control. Small molecules (antibiotics, antivirals, plant-derived secondary molecules, anticancer agents, or synthetic compounds) can provide this transcriptional control to regulate DNA modifications, transcriptional regulation, posttranscriptional controlling, protein trafficking, and posttranslational regulation. On the other hand, using approved drugs or food ingredients like gluconate or xylene to activate genetic circuits provides advantages at the clinical level of treatment strategies.

Gluconate is a metabolic product in different cell types, including *E. coli*, but it is also a carbon source. In some circumstances, *E. coli* can be controlled by transcriptional repressor gluconate-inducible promoter GntR. A recently published study used this strategy to control blood sugar levels in diabetic mice. A designed gluconate-induced genetic switch was linked to the insulin gene to provide a balanced blood glucose level in diabetic mice to reach normal blood sugar levels, and auspicious results are taken in this concept [81]. The same strategy was applied by the same research group for xylose [82] and acetoin [83] depending on genetic regulation. Both systems have proven that transcription factor-mediated mechanisms can be used for in situ therapeutical gene regulation using a similar approach. Usage of antibiotic and antiviral molecules to inhibit or activate different gene cascades is another useful approach to regulating cellular systems. One of the first applications of this approach was the development of coumermycin sensors [84].

Disease monitoring is an essential application for precision medicine and the first stage of treatment with modified microorganisms. These versatile and robust biologically controlled genetic circuits can be integrable to the organisms that are located in specific targets such as the intestinal lumen. A recent study applied this strategy to monitor gastrointestinal health with heme-sensitive probiotic biosensors to detect intestinal bleeding. Sensor activation starts with luminescence protein activation in the intestinal area, which can be detected quickly with wireless sensors [85].

2.3.2.1.4 Heat-inducible systems

Thermosensitive gene regulation systems are natural systems that have evolved to show resistance to environmental changes. This regulation mechanism is important because of its sensitivity and easy controllability. These systems can also be modified using synthetic biology applications for therapeutic purposes. Regulation can be provided with temperature-dependent conformation-controlled transcription factors, which can be adapted to different organisms with various modifications. As a successful example of these applications, Tcl proteins are a frequently used protein expression regulator for transcription control because of the high metabolism rate depending on increased temperatures in the tumor area. The transactivation capacity of Tcl family proteins can provide strict transcription regulation in tuned conditions. Xiong et al. [86] developed a genetic circuit controlled by body temperature to repress or activate targeted gene expression. The applicability of these regulators to various organisms also provides an advantage in the applications. The modified transitive transcription systems have been reported in the literature recently; a synthetic membrane potential mediated with engineered TRPV1-mediated calcium influx transcription regulation was developed, which can be activated by sunlight or capsaicin to treat diabetic mice. Both activators provide heat on the transcription site as a starting factor of transcription [87]. Using heat-activated gene regulation systems offers many advantages but must also be considered for uncontrolled temperature changes. Because of that, in situ therapeutic applications of this controlling system have limits on application level.

2.3.2.1.5 Electric stimulated systems

Electrical stimulation is another mechanism for controlled therapeutical release systems, which can provide different advantages, especially for some cell types and during application. Krawczyk and his colleagues have developed a cellular system based on a membrane depolarization-mediated controlling system that stimulates a wireless electric signal. For this purpose, they modified calcium-mediated voltage channels, which allow real-time control of the release of protein of interest. With the designed genetic circuit, human beta cells are engineered for electrosensitivity and tested subcutaneously through injection of mice. The programmed cells have released vesicular insulin via wireless electrical stimulation, which has been provided to reach normoglycemia in type 1 diabetes [88].

2.3.2.1.6 Ultrasound stimulated systems

Sound-controlled systems are widely used in material science, chemistry, and engineering; however they can also be applied to transcription control with synthetic biology tools. The specific wavelengths can stimulate association and disassociation processes, which are adaptable to biological organisms to control transcription. Usage of *S. typhimurium* seems a safe and effective tool for cargo carrying in the clinical approaches, and different studies in the clinical phase still continue. In a recent study, various Tcl transcriptional repressor-TlpA39 controlled *S. typhimurium* cells were directed to programmed death-ligand 1 (PD L1), representing a target for settling until ultrasound activation. This sonogenetic-controlled therapeutical concept has shown promising results because of well-controlled transcription in the localized regions [89].

2.4 Conclusion

In conclusion, engineered biological devices via synthetic gene circuits offer precision in controlling cellular functions and responses that provide advancement in the field of synthetic biology and biotechnology. Cancer cells can be diagnosed and treated with high specificity by engineering receptors and signaling pathways. Gene expression can be precisely controlled by light-inducible and heat-responsive systems to improve the therapeutic behavior of machineries. The combinatorial effect of logic gates and synthetic gene networks enables the multiple signal integration and production of coordinated output. The safety and efficacy of treatment can be improved with the designed biological devices.

References

[1]   Bush LM, Healy CP, Javdan SB, Emmons JC, Deans TL. Biological cells as therapeutic delivery vehicles. Trends Pharmacol Sci 2021;42:106. https://doi.org/10.1016/J.TIPS.2020.11.008.

[2]   Frejd FY, Kim KT. Affibody molecules as engineered protein drugs. Exp Mol Med 2017;49(3):e306–e306. https://doi.org/10.1038/emm.2017.35.

[3]   Gujrati V, Kim S, Kim SH, Min JJ, Choy HE, Kim SC, et al. Bioengineered bacterial outer membrane vesicles as cell-specific drug-delivery vehicles for cancer therapy. ACS Nano 2014;8:1525–37. https://doi.org/10.1021/NN405724X.

[4]   Slomovic S, Pardee K, Collins JJ. Synthetic biology devices for in vitro and in vivo diagnostics. Proc Natl Acad Sci USA 2015;112:14429–35. https://doi.org/10.1073/PNAS.1508521112/ASSET/4DBE653E-8604-42A8-8C71-CA6E5838A898/ASSETS/GRAPHIC/PNAS.1508521112FIG02.JPEG.

[5]   Douglas SM, Bachelet I, Church GM. A logic-gated nanorobot for targeted transport of molecular payloads. Science (80-) 2012;335:831–34. https://doi.org/10.1126/SCIENCE.1214081/SUPPL_FILE/DOUGLAS.SOM-CORRECTED.PDF.

[6]   Wu Y, Liu Y, Huang Z, Wang X, Jin Z, Li J, et al. Control of the activity of CAR-T cells within tumours via focused ultrasound. Nat Biomed Eng 2021;5(11):1336–47. https://doi.org/10.1038/s41551-021-00779-w.

[7]   Miller IC, Zamat A, Sun LK, Phuengkham H, Harris AM, Gamboa L, et al. Enhanced intratumoural activity of CAR T cells engineered to produce immunomodulators under photothermal control. Nat Biomed Eng 2021;5(11):1348–59. https://doi.org/10.1038/s41551-021-00781-2.

[8]   Huang Z, Wu Y, Allen ME, Pan Y, Kyriakakis P, Lu S, et al. Engineering light-controllable CAR T cells for cancer immunotherapy. Sci Adv 2020;6. https://doi.org/10.1126/SCIADV.AAY9209/SUPPL_FILE/AAY9209_TABLE_S1.XLSX.

[9]   Lim B, Yin Y, Ye H, Cui Z, Papachristodoulou A, Huang WE. Reprogramming synthetic cells for targeted cancer therapy. ACS Synth Biol 2022;11:1349–60. https://doi.org/10.1021/ACSSYNBIO.1C00631/ASSET/IMAGES/MEDIUM/SB1C00631_M001.GIF.

[10]   Zhao N, Song Y, Xie X, Zhu Z, Duan C, Nong C, et al. Synthetic biology-inspired cell engineering in diagnosis, treatment, and drug development. Signal Transduct Target Ther 2023;8(1):1–21. https://doi.org/10.1038/s41392-023-01375-x.

[11]   Courbet A, Endy D, Renard E, Molina F, Bonnet J. Detection of pathological biomarkers in human clinical samples via amplifying genetic switches and logic gates. Sci Transl Med 2015;7. https://doi.org/10.1126/scitranslmed.aaa3601.

[12]   Park SJ, Park SH, Cho S, Kim DM, Lee Y, Ko SY, et al. New paradigm for tumor theranostic methodology using bacteria-based microrobot. Sci Rep 2013;3. https://doi.org/10.1038/SREP03394.

[13]   Mao N, Cubillos-Ruiz A, Cameron DE, Collins JJ. Probiotic strains detect and suppress cholera in mice. Sci Transl Med 2018;10:0–9. https://doi.org/10.1126/scitranslmed.aao2586.

[14]   Din MO, Danino T, Prindle A, Skalak M, Selimkhanov J, Allen K, et al. Synchronized cycles of bacterial lysis for in vivo delivery. Nature 2016;536:81–85. https://doi.org/10.1038/nature18930.

[15]   Chen X, Li P, Xie S, Yang X, Luo B, Hu J. Genetically engineered probiotics for optical imaging-guided tumor photothermal /immunotherapy. Biomater Sci 2023. https://doi.org/10.1039/d3bm01227a.

[16]   Wang W, Xu H, Ye Q, Tao F, Wheeldon I, Yuan A, et al. Systemic immune responses to irradiated tumours via the transport of antigens to the tumour periphery by injected flagellate bacteria. Nat Biomed Eng 2022;6(1):44–53. https://doi.org/10.1038/s41551-021-00834-6.

[17]   Hu Q, Wu M, Fang C, Cheng C, Zhao M, Fang W, et al. Engineering nanoparticle-coated bacteria as oral DNA vaccines for cancer immunotherapy. Nano Lett 2015;15:2732–39. https://doi.org/10.1021/ACS.NANOLETT.5B00570/SUPPL_FILE/NL5B00570_SI_001.PDF.

[18]   Balázsi G, van Oudenaarden A, Collins JJ. Cellular decision-making and biological noise: From microbes to mammals. Cell 2011;144:910–25. https://doi.org/10.1016/j.cell.2011.01.030.

[19]   Kerr R, Jabbari S, Johnston IG. Intracellular energy variability modulates cellular decision-making capacity. Sci Rep 2019;9:20196. https://doi.org/10.1038/s41598-019-56587-5.

[20]   Kobayashi TJ, Kamimura A. Theoretical Aspects of Cellular Decision-Making and Information-Processing. In: Goryanin II, Goryachev AB, editors. Adv. Syst. Biol., vol. 736, New York, NY: Springer New York; 2012, pp. 275–91.

[21]   Perkins TJ, Swain PS. Strategies for cellular decision-making. Mol Syst Biol 2009;5:326. https://doi.org/10.1038/msb.2009.83.

[22]   Balaban NQ, Merrin J, Chait R, Kowalik L, Leibler S. Bacterial persistence as a phenotypic switch. Science (80-) 2004;305:1622–25. https://doi.org/10.1126/science.1099390.

[23]   Mukherjee S, Bassler BL. Bacterial quorum sensing in complex and dynamically changing environments. Nat Rev Microbiol 2019;17:371–82. https://doi.org/10.1038/s41579-019-0186-5.

[24]   Papenfort K, Bassler BL. Quorum sensing signal–response systems in Gram-negative bacteria. Nat Rev Microbiol 2016;14:576–88. https://doi.org/10.1038/nrmicro.2016.89.

[25]   Moreno-Gámez S, Hochberg ME, van Doorn GS. Quorum sensing as a mechanism to harness the wisdom of the crowds. Nat Commun 2023;14:3415. https://doi.org/10.1038/s41467-023-37950-7.

[26] You L, Cox RS, Weiss R, Arnold FH. Programmed population control by cell–cell communication and regulated killing. Nature 2004;428:868–71. https://doi.org/10.1038/nature02491.

[27] Shahbazi MN. Mechanisms of human embryo development: From cell fate to tissue shape and back. Development 2020;147:dev190629. https://doi.org/10.1242/dev.190629.

[28] Takahashi K, Tanabe K, Ohnuki M, Narita M, Ichisaka T, Tomoda K, et al. Induction of pluripotent stem cells from adult human fibroblasts by defined factors. Cell 2007;131:861–72. https://doi.org/10.1016/j.cell.2007.11.019.

[29] Mateos-Aparicio P, Rodríguez-Moreno A. The impact of studying brain plasticity. Front Cell Neurosci 2019;13:66. https://doi.org/10.3389/fncel.2019.00066.

[30] Cookson S, Ostroff N, Pang WL, Volfson D, Hasty J. Monitoring dynamics of single-cell gene expression over multiple cell cycles. Mol Syst Biol 2005;1:2005–24. https://doi.org/10.1038/msb4100032.

[31] Groisman A, Lobo C, Cho H, Campbell JK, Dufour YS, Stevens AM, et al. A microfluidic chemostat for experiments with bacterial and yeast cells. Nat Methods 2005;2:685–89. https://doi.org/10.1038/nmeth784.

[32] Khalil AS, Collins JJ. Synthetic biology: Applications come of age. Nat Rev Genet 2010;11:367–79. https://doi.org/10.1038/nrg2775.

[33] Lu TK, Collins JJ. Engineered bacteriophage targeting gene networks as adjuvants for antibiotic therapy. Proc Natl Acad Sci 2009;106:4629–34. https://doi.org/10.1073/pnas.0800442106.

[34] Anderson JC, Clarke EJ, Arkin AP, Voigt CA. Environmentally controlled invasion of cancer cells by engineered bacteria. J Mol Biol 2006;355:619–27. https://doi.org/10.1016/j.jmb.2005.10.076.

[35] Nissim L, Wu M-R, Pery E, Binder-Nissim A, Suzuki HI, Stupp D, et al. Synthetic RNA-based immunomodulatory gene circuits for cancer immunotherapy. Cell 2017;171:1138–50, e15. https://doi.org/10.1016/j.cell.2017.09.049.

[36] Aijaz A, Li M, Smith D, Khong D, LeBlon C, Fenton OS, et al. Biomanufacturing for clinically advanced cell therapies. Nat Biomed Eng 2018;2:362–76. https://doi.org/10.1038/s41551-018-0246-6.

[37] Jacob F, Monod J. Genetic regulatory mechanisms in the synthesis of proteins. J Mol Biol 1961;3:318–56. https://doi.org/10.1016/S0022-2836(61)80072-7.

[38] Thomas R. Boolean formalization of genetic control circuits. J Theor Biol 1973;42.

[39] Hınçer A, Ahan RE, Aras E, Şeker UÖŞ. Making the next generation of therapeutics: MRNA meets synthetic biology. ACS Synth Biol 2023;12:2505–15. https://doi.org/10.1021/acssynbio.3c00253.

[40] Brázda V, Bartas M, Bowater RP. Evolution of diverse strategies for promoter regulation. Trends Genet 2021;37:730–44. https://doi.org/10.1016/j.tig.2021.04.003.

[41] Andersson R, Sandelin A. Determinants of enhancer and promoter activities of regulatory elements. Nat Rev Genet 2020;21:71–87. https://doi.org/10.1038/s41576-019-0173-8.

[42] Ni X, Liu Z, Guo J, Zhang G. Development of terminator–promoter bifunctional elements for application in *Saccharomyces cerevisiae* pathway engineering. Int J Mol Sci 2023;24. https://doi.org/10.3390/ijms24129870.

[43] Wang Y, Wang H, Wei L, Li S, Liu L, Wang X. Synthetic promoter design in Escherichia coli based on a deep generative network. Nucleic Acids Res 2020;48:6403–12. https://doi.org/10.1093/nar/gkaa325.

[44] Patikoglou G, Burley SK. Eukaryotic transcription factor-DNA complexes. Annu Rev Biophys Biomol Struct 1997;26.

[45] Lawson CL, Swigon D, Murakami KS, Darst SA, Berman HM, Ebright RH. Catabolite activator protein: DNA binding and transcription activation. Curr Opin Struct Biol 2004;14:10–20. https://doi.org/10.1016/j.sbi.2004.01.012.

[46] Mandal PK, Collie GW, Srivastava SC, Kauffmann B, Huc I. Structure elucidation of the Pribnow box consensus promoter sequence by racemic DNA crystallography. Nucleic Acids Res 2016;44:5936–43. https://doi.org/10.1093/nar/gkw367.

[47] Field A, Adelman K. Evaluating enhancer function and transcription. 2020. https://doi.org/10.1146/annurev-biochem-011420.

[48] Clare SE, Hatfield WR, Fantz DA, Kistler WS, Kistler MK. Characterization of the promoter region of the rat testis-specific histone Hlt gene'. Biol Reprod 1997;56.

[49] Rossi JJ, Soberon X, Marumoto Y, Mcmahon J. Biological expression of an Escherichia coli consensus sequence promoter and some mutant derivatives (synthetic DNA/mutagenesis/heparin resistance). Proc Natl Acad Sci 1983;80.

[50] Blow MJ, Clark TA, Daum CG, Deutschbauer AM, Fomenkov A, Fries R, et al. The epigenomic landscape of prokaryotes. PLoS Genet 2016;12. https://doi.org/10.1371/journal.pgen.1005854.

[51] Lee TF, Zhai J, Meyers BC. Conservation and divergence in eukaryotic DNA methylation. Proc Natl Acad Sci USA 2010;107:9027–28. https://doi.org/10.1073/pnas.1005440107.

[52] Huang L, Yuan Z, Liu P, Zhou T. Effects of promoter leakage on dynamics of gene expression. BMC Syst Biol 2015;9. https://doi.org/10.1186/s12918-015-0157-z.

[53] Heiss S, Hörmann A, Tauer C, Sonnleitner M, Egger E, Grabherr R, et al. Evaluation of novel inducible promoter/repressor systems for recombinant protein expression in Lactobacillus plantarum. Microb Cell Fact 2016;15. https://doi.org/10.1186/s12934-016-0448-0.

[54] Ali MZ, Brewster RC. Controlling gene expression timing through gene regulatory architecture. PLoS Comput Biol 2022;18. https://doi.org/10.1371/journal.pcbi.1009745.

[55] Ermolenko DN, Mathews DH. Making ends meet: New functions of mRNA secondary structure. Wiley Interdiscip Rev RNA 2021;12. https://doi.org/10.1002/wrna.1611.

[56] Guntas G, Hallett RA, Zimmerman SP, Williams T, Yumerefendi H, Bear JE, et al. Engineering an improved light-induced dimer (iLID) for controlling the localization and activity of signaling proteins. Proc Natl Acad Sci USA 2015;112:112–17. https://doi.org/10.1073/pnas.1417910112.

[57] Takala H, Björling A, Linna M, Westenhoff S, Ihalainen JA. Light-induced changes in the dimerization interface of bacteriophytochromes. J Biol Chem 2015;290:16383–92. https://doi.org/10.1074/jbc.M115.650127.

[58] Kallunki T, Barisic M, Jäättelä M, Liu B. How to choose the right inducible gene expression system for mammalian studies? Cells 2019;8. https://doi.org/10.3390/cells8080796.

[59] He S, Zhang Z, Lu W. Natural promoters and promoter engineering strategies for metabolic regulation in *Saccharomyces cerevisiae*. J Ind Microbiol Biotechnol 2023;50. https://doi.org/10.1093/jimb/kuac029.

[60] Nistala GJ, Wu K, Rao CV, Bhalerao KD. A modular positive feedback-based gene amplifier. J Biol Eng 2010;4.

[61] Yamada M, Nagasaki SC, Ozawa T, Imayoshi I. Light-mediated control of Gene expression in mammalian cells. Neurosci Res 2020;152:66–77. https://doi.org/10.1016/j.neures.2019.12.018.

[62] Ochoa-Fernandez R, Abel NB, Wieland FG, Schlegel J, Koch LA, Miller JB, et al. Optogenetic control of gene expression in plants in the presence of ambient white light. Nat Methods 2020;17:717–25. https://doi.org/10.1038/s41592-020-0868-y.

[63] Sheets MB, Wong WW, Dunlop MJ. Light-inducible recombinases for bacterial optogenetics. ACS Synth Biol 2020;9:227–35. https://doi.org/10.1021/acssynbio.9b00395.

[64] Jung JC, Schnitzer MJ. Multiphoton endoscopy. Opt Lett 2003;28.

[65] Ruess J, Parise F, Milias-Argeitis A, Khammash M, Lygeros J. Iterative experiment design guides the characterization of a light-inducible gene expression circuit. Proc Natl Acad Sci U S A 2015;112:8148–53. https://doi.org/10.1073/pnas.1423947112.

[66] Raghavan AR, Salim K, Yadav VG. Optogenetic control of heterologous metabolism in E. coli. ACS Synth Biol 2020;9:2291–300. https://doi.org/10.1021/acssynbio.9b00454.

[67] Kaberniuk AA, Baloban M, Monakhov MV, Shcherbakova DM, Verkhusha VV. Single-component near-infrared optogenetic systems for gene transcription regulation. Nat Commun 2021;12. https://doi.org/10.1038/s41467-021-24212-7.

[68] Zhou Y, Kong D, Wang X, Yu G, Wu X, Guan N, et al. A small and highly sensitive red/far-red optogenetic switch for applications in mammals. Nat Biotechnol 2022;40:262–72. https://doi.org/10.1038/s41587-021-01036-w.

[69] Zhu X, Chen S, Hu X, Zhao L, Wang Y, Huang J, et al. Near-infrared nano-optogenetic activation of cancer immunotherapy via engineered bacteria. Adv Mater 2023;35. https://doi.org/10.1002/adma.202207198.

[70] Cui M, Pang G, Zhang T, Sun T, Zhang L, Kang R, et al. Optotheranostic nanosystem with phone visual diagnosis and optogenetic microbial therapy for ulcerative colitis at-home care. ACS Nano 2021;15:7040–52. https://doi.org/10.1021/acsnano.1c00135.

[71] Mansouri M, Hussherr MD, Strittmatter T, Buchmann P, Xue S, Camenisch G, et al. Smart-watch-programmed green-light-operated percutaneous control of therapeutic transgenes. Nat Commun 2021;12. https://doi.org/10.1038/s41467-021-23572-4.

[72] Mansouri M, Xue S, Hussherr MD, Strittmatter T, Camenisch G, Fussenegger M. Smartphone-flashlight-mediated remote control of rapid insulin secretion restores glucose homeostasis in experimental type-1 diabetes. Small 2021;17. https://doi.org/10.1002/smll.202101939.

[73] Nihongaki Y, Yamamoto S, Kawano F, Suzuki H, Sato M. CRISPR-Cas9-based photoactivatable transcription system. Chem Biol 2015;22:169–74. https://doi.org/10.1016/j.chembiol.2014.12.011.

[74] Zhou XX, Zou X, Chung HK, Gao Y, Liu Y, Qi LS, et al. A single-chain photoswitchable CRISPR-Cas9 architecture for light-inducible gene editing and transcription. ACS Chem Biol 2018;13:443–48. https://doi.org/10.1021/acschembio.7b00603.

[75] Hu M, Liu R, Qiu Z, Cao F, Tian T, Lu Y, et al. Light-Start CRISPR-Cas12a reaction with caged crRNA enables rapid and sensitive nucleic acid detection. Angew Chemie – Int Ed 2023;62. https://doi.org/10.1002/anie.202300663.

[76] Leng K, Kampmann M. Towards elucidating disease-relevant states of neurons and glia by CRISPR-based functional genomics. Genome Med 2022;14. https://doi.org/10.1186/s13073-022-01134-7.

[77] Shao J, Wang M, Yu G, Zhu S, Yu Y, Heng BC, et al. Synthetic far-red light-mediated CRISPR-dCas9 device for inducing functional neuronal differentiation. Proc Natl Acad Sci USA 2018;115:E6722–30. https://doi.org/10.1073/pnas.1802448115.

[78] Allen ME, Zhou W, Thangaraj J, Kyriakakis P, Wu Y, Huang Z, et al. An AND-gated drug and photoactivatable Cre-loxP system for spatiotemporal control in cell-based therapeutics. ACS Synth Biol 2019;8:2359–71. https://doi.org/10.1021/acssynbio.9b00175.

[79] Zhong G, Wang H, He W, Li Y, Mou H, Tickner ZJ, et al. A reversible RNA on-switch that controls gene expression of AAV-delivered therapeutics in vivo. Nat Biotechnol 2020;38:169–75. https://doi.org/10.1038/s41587-019-0357-y.

[80] Lajoie MJ, Boyken SE, Salter AI, Bruffey J, Rajan A, Langan RA, et al. Designed protein logic to target cells with precise combinations of surface antigens. Science n.d.

[81] Teixeira AP, Xue S, Huang J, Fussenegger M. Evolution of molecular switches for regulation of transgene expression by clinically licensed gluconate. Nucleic Acids Res 2023;51:E85–E85. https://doi.org/10.1093/nar/gkad600.

[82] Galvan S, Madderson O, Xue S, Teixeira AP, Fussenegger M. Regulation of transgene expression by the natural sweetener xylose. Adv Sci 2022;9. https://doi.org/10.1002/advs.202203193.

[83] Bertschi A, Stefanov BA, Xue S, Charpin-El Hamri G, Teixeira AP, Fussenegger M. Controlling therapeutic protein expression via inhalation of a butter flavor molecule. Nucleic Acids Res 2023;51. https://doi.org/10.1093/nar/gkac1256.

[84] Zhao H-F, Boyd J, Jolicoeur N, Shen S-H. A coumermycin/novobiocin-regulated gene expression system. Hum Gene Ther n.d.

[85] Mimee M, Nadeau P, Hayward A, Carim S, Flanagan S, Jerger L, et al. An ingestible bacterial-electronic system to monitor gastrointestinal health. Science (80-) 2018;360:915–18. https://doi.org/10.1126/science.aas9315.

[86] Xiong LL, Garrett MA, Buss MT, Kornfield JA, Shapiro MG. Tunable temperature-sensitive transcriptional activation based on lambda repressor. ACS Synth Biol 2022;11:2518–22. https://doi.org/10.1021/acssynbio.2c00093.

[87] Stefanov BA, Mansouri M, Charpin-El Hamri G, Fussenegger M. Sunlight-controllable biopharmaceutical production for remote emergency supply of directly injectable therapeutic proteins. Small 2022;18. https://doi.org/10.1002/smll.202202566.

[88] Krawczyk K, Xue S, Buchmann P, Charpin-El-Hamri G, Saxena P, Hussherr M-D, et al. Electrogenetic cellular insulin release for real-time glycemic control in type 1 diabetic mice. Science n.d.

[89] Gao T, Niu L, Wu X, Dai D, Zhou Y, Liu M, et al. Sonogenetics-controlled synthetic designer cells for cancer therapy in tumor mouse models. Cell Reports Med 2024. https://doi.org/10.1016/j.xcrm.2024.101513.

Melis Karaca, Senem Şen, Fayiti Fuerkaiti,
and Urartu Özgür Şafak Şeker

Chapter 3
Engineering mammalian cell for cancer

Abstract: The accuracy of products produced by mammalian cells can be explained by the fact that continuously stored environmental information leads to high-precision production capacity. Therefore, usage of mammalian cells in cancer might lead to progression of safe and efficient therapies. Synthetic biology has emerged as a transformative force in biomedical innovation, which aims to reprogram cells for precise diagnostics and therapeutics. Recent advances highlight the development of modular synthetic receptors, enabling customizable disease recognition and engineered mammalian cells exhibiting remarkable sensitivity and selectivity. Simultaneously, advances in mammalian synthetic biology facilitate the deliberate engineering of protein secretion, glycosylation, cellular metabolism, and cellular communication, unlocking new therapeutic possibilities that even lead to the construction of artificial tissues for innovative cancer therapies. This chapter focuses on the current techniques of mammalian cell engineering for cancer therapeutics including their drawbacks and future.

3.1 Introduction

Mammalian cells are eukaryotic cells. These are essential units of life in multicellular organisms, which play vital roles in various biological processes. The human body consists of trillions of individual cells in tissues and organs. These cells have a complex structure that enables them to adapt and specialize for specific functions. As a result, mammalian cells in the human body encompass a wide array of cell types carrying out various physiological functions. Therefore, understanding the mechanisms underlying the mammalian cell's various biological functions is essential [1].

Cancer can be defined as a complex and diverse group of diseases characterized by uncontrolled growth, survival, and, in some cases, the spread of abnormal cells in the body through metastasis [2]. Cancer is a genetic disease. Therefore, disruptions in mammalian cells' normal genetic regulatory mechanisms can lead to uncontrolled cell growth and proliferation, resulting in cancer progression [3]. As a result of the increased life span, the reoccurrence of cancer increases every year. By 2050, around 35 million people are estimated to have cancer annually, which indicates a 77% in-

Melis Karaca, Senem Şen, Fayiti Fuerkaiti, Urartu Özgür Şafak Şeker, UNAM–Institute of Materials Science and Nanotechnology, Bilkent University

https://doi.org/10.1515/9783111329499-003

crease compared to the 2022 level. It encompasses various types and subtypes, each with distinct characteristics and behaviors that can be lethal. It is responsible for 16.8-22.8% of noncommunicable diseases in the twenty-first century. Based on the death records in 2022, the most lethal cancer type is lung cancer, followed by breast cancer. It is known that these numbers vary between genders and age ranges [4].

Evidence indicates that a series of genetic mutations in a single cell could lead to cancer [3]. Advancements in understanding the molecular genetics of cancer revealed the genetic basis of various types of cancer. For instance, mutations in critical genetic regulators of hematopoiesis have been associated with hematopoietic neoplasms, highlighting the link between genetic alterations and cancer development [5]. Furthermore, the dysregulation of transcription factors like RUNX1 (RUNX family transcription factor 1) has been implicated in leukemia, emphasizing the importance of proper gene expression regulation in preventing oncogenesis [6]. It is possible to tailor novel treatments based on mammalian cells by elucidating the relationship between genetic mutation and cancer development. In recent timelines, synthetic gene circuits in mammalian cells have shown promise in cancer therapy [7].

Synthetic biology is a multidisciplinary field that utilizes engineering principles to design systems for producing biological molecules, sensors, and systems. With synthetic biology, it is possible to create cell factories that provide the needs of industry or medicine from renewable sources. It allows the creation of highly specific and tunable cells for various functions. The use of these cells in different areas, such as agriculture, medicine, and industry, is increasing with the advancements in the field. For cancer, synthetic biology holds great promise since the optimized systems can be used for biomanufacturing, diagnosis, and therapy. In this chapter, the use of these systems for cancer will be given, and issues regarding engineering protein secretion, glycosylation, cellular communication, and metabolism will be discussed [1].

3.2 Engineering protein secretion

Proteins designated for secretion or residence in the plasma membrane pass through interconnected steps within the endoplasmic reticulum (ER) and Golgi apparatus. This protracted journey involves vesicle intermediates emanating from a donor compartment and fusing with an acceptor/target compartment. The adjustment of vesicle budding and cargo selection in the donor compartment involves coat protein complexes (COP) such as COPI (coat protein complex I), COPII (coat protein complex II), and clathrin. In contrast, vesicle fusion relies on the interplay between SNARE (soluble *N*-ethylmaleimide-sensitive factor attachment protein receptor) and Sec1/Munc18 proteins. The secretory pathway, primarily unfolding in the ER, Golgi apparatus, and the endomembrane system, holds particular significance in biotechnology and the biopharmaceutical industry. Notably, the entry of proteins with *N*-terminal signal sequences into the ER represents the initial step in the secretory pathway, with pre-proteins gaining access either cotranslationally or posttranslationally. This complex process involves the signal sequence recognition particle (SRP). After synthesis on the ribosome-bound ER, secretory proteins undergo translocation into the ER lumen via a cotranslational mechanism. Transport vesicles facilitate the departure of secretory proteins from the ER, with vesicle budding mediated by COPs. Tethering and fusion of vesicles, crucial for cargo unloading, involve multicomponent heteromeric protein complexes (tethering factors) and evolutionarily conserved Rab GTPases (GTP-binding proteins). The secretion process is further characterized by two modes: constitutive, wherein proteins are secreted concurrently with synthesis, and regulated, where secretory proteins are stored in transport vesicles before release. Protein folding within the ER involves the formation of disulfide bonds, prolyl isomerization, and the sequestration of hydrophobic amino acids. The complexities of these processes underline the significance of comprehensive understanding and strategic engineering for improved protein secretion in mammalian cells. The secretory pathway encompasses complex processes predominantly occurring within the ER, Golgi apparatus, and the endomembrane system. Facilitating the proper targeting of newly synthesized proteins to their ultimate destinations is a crucial function of the secretory pathway, thereby upholding the cell structure and function. This pathway starts with entering proteins with *N*-terminal signal sequences into the ER, a process that can occur either cotranslationally or posttranslationally [8–10].

In biopharmaceutical manufacturing, the majority of products are secreted proteins. However, the limited secretion capacity of mammalian cells poses a significant challenge. Advancing our understanding of vesicle trafficking and developing engineering strategies to enhance secretion are crucial for maximizing production potential [11].

3.2.1 Signal peptides and optimization of signal peptides for enhanced secretion

Signal peptides, short sequences located at the *N*-terminal region of proteins, serve as directives for protein secretion. The initial phase of protein secretion relies on signal peptides, guiding the association of the translating ribosome with the SRP and facilitating translation arrest. While various signal peptide sequences show differing efficiencies in promoting the secretion of heterologous proteins, those from interleukins or immunoglobulins (Igs) are commonly used for directing recombinant protein secretion. However, challenges arise in selecting an appropriate signal sequence due to variations associated with host cells, secretion processes, and interactions with subsequent portions of the polypeptide [12, 13].

Bioengineering host cell lines can enhance the modification or secretion of heterologous proteins. A comprehensive understanding of these enhancements' molecular mechanisms and their particular effects on diverse recombinant proteins requires systematic exploration [12, 13]. In this sense, literature following the methods developed in the literature is possible. In line with this, Kober et al. [14] focused on optimizing signal peptides to increase the secretory efficiency of recombinant biotherapeutic antibodies [14]. They identified signal peptides B and E, which they named in their study, as promising candidates for improving production rates suitable for commercial use. However, in addition to known signal peptides, a very large pool can be created to recognize signal peptides, and potential candidates can be selected and developed from these pools. Additionally, the research by Park et al. (2022) addressed the difficulties in developing an in vitro screening system for signal peptides in mammalian cells [15]. By combining degenerate codon-based oligonucleotide library construction, an Flp-In™ system, and a fluorescence-activated cell sorting-based cold capture assay, they created a screening system to characterize single clones transfected with signal peptides. Furthermore, studies have continued to support the statement of Bachhav et al. (2023), which stated that the importance of signal peptides in the secretion of therapeutic proteins and their recognition by SRP during translocation into the ER is crucial [16]. As a result, Cheng et al. [17] reported the effect of artificial signal peptides on the secretion activity of recombinant lysosomal enzymes in CHO (Chinese hamster ovarian) cell lines [17]. They engineered two recombinant human lysosomal enzymes, N-acetyl-α-glucosaminidase (rhNAGLU) and glucosamine (N-acetyl)-6-sulfatase (rhGNS), by replacing their native signal peptides with nine different signal peptides derived from highly secretory proteins in CHO K1 cells. When comparing native signal peptides, rhGNS has been noted to be secreted into media at higher levels than rhNAGLU. It is concluded that the secretion of rhNAGLU and rhGNS can be carefully controlled by changing the signal peptides. Finally, the existing computational tools for predicting unconventional protein secretion (UPS), such as SecretomeP and SPRED, are products of a hypothesis based on which all secretory proteins share common properties, independent of specific pathways. Focusing on this, it exploits classical secretory proteins by removing signal peptides. In another work, Zhao et al. (2019) introduced OutCyte, a new tool for UPS prediction that covers secretory pathways that do not contain a defined N-terminal signal peptide [18]. OutCyte-UPS outperformed existing tools in predicting UPS proteins, laying the foundation for improved UPS prediction in the future using experimentally validated data.

3.2.2 Cell line development to improve secretion efficiency

Cell line development is included in the drug development process, specifically at the stage following the identification of a potential drug as a protein. Cell line engineering is also included, especially in protein secretion enhancement studies, genetic circuit designs, and cleavage of ER-retention signal sequence studies. In pursuit of precise control over protein secretion, innovative platforms have been developed. Vlahos et al. [19] introduced RELEASE and RELEASE-NOT systems (Figure 3.1) to modulate in-

A)

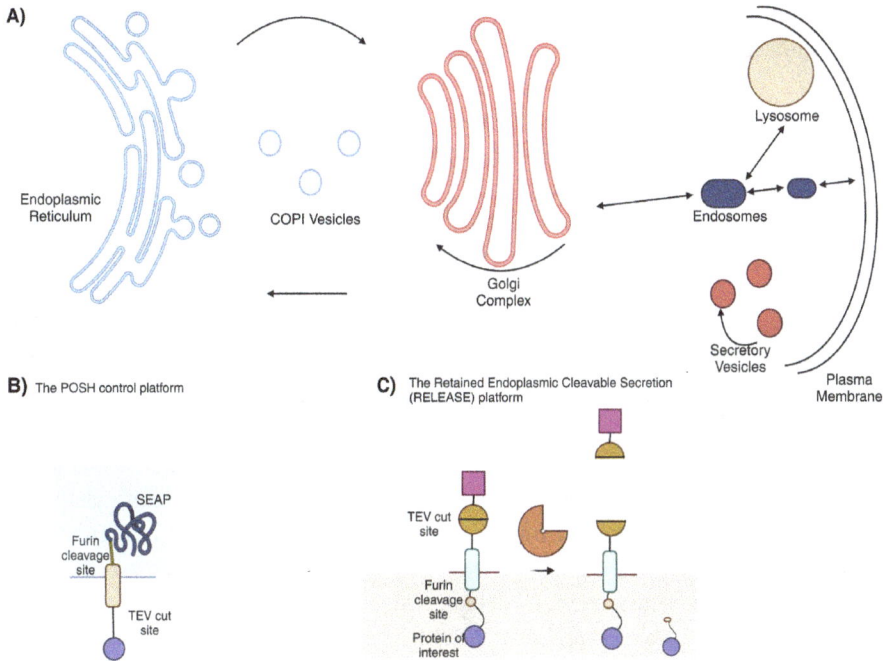

Endoplasmic
Reticulum

COPI Vesicles

Golgi
Complex

Lysosome

Endosomes

Secretory
Vesicles

Plasma
Membrane

B) The POSH control platform

SEAP

Furin
cleavage
site

TEV cut
site

C) The Retained Endoplasmic Cleavable Secretion
(RELEASE) platform

TEV cut
site

Furin
cleavage
site

Protein of
interest

Figure 3.1: Schematic representation of (A) the engineerable protein trafficking network and secretion pathway in mammalian cells and (B) the POSH platform [20] that uses an inducible protease to remove the ER import signal from a synthetic reporter, allowing it to move to the trans-Golgi region, where it is cleaved by furin for secretion. (C) The RELEASE platform employs ER retention motifs to protect tagged proteins in the ER, which are then released through protease activation and transported via the conventional secretory pathway [19].

tercellular signaling through proteolytic removal of ER-retention motifs, paving the way for complex expression profiles. The RELEASE platform was modified to enable negative regulation of protein secretion (RELEASE-NOT) in response to protease activity. This advancement allows for tasks like logical processing and filtering of analog signals using just one protease input [19]. Combining these into compRELEASE, the authors achieved logic processing and signal filtering with a single protease input. In addition, Mansouri et al. [20] designed the programmable protease-mediated post-translational switch (POSH) platform, demonstrating efficient protein secretion with protease-driven protein circuits. This approach generates the desired protein in a chimeric form linked to an ER-retention signal (Lys-Lys-Tyr-Leu (KKYL)), keeping the protein within the ER under normal conditions. When triggered by an inducer, two fragmented cytoplasmic protease units merge to create an active protease. This activated protease removes the ER-retention signal, initiating the release and secretion of the protein stored in the ER [20]. Praznik et al. [21] utilized split potyviral proteases for the controlled release of ER-lumen proteins, offering a faster alternative to transcription-based systems [21].

3.2.3 Manipulating vesicle transport for improved protein secretion

Intracellular vesicle trafficking is critical in maintaining organelle homeostasis in eukaryotic cells. Studies suggesting pharmacological manipulation of membrane fusion as a potential therapeutic approach are important [22]. In unusual protein secretion mechanisms that play an important role in releasing proteins outside the cell, this process occurs through extracellular vesicles (EVs), including microvesicles and exosomes. Exosomes are formed within endocytic vacuoles and are released by fusion with lysosomes or exocytosis. Ectosomes are released directly from the plasma membrane [23]. Interventions at the cellular level targeting the translocation machinery have proven effective in enhancing recombinant protein production in mammalian cells.

For instance, Le Fourn et al. [13] demonstrated increased translocation of proteins into the ER lumen by overexpressing SRP14 (signal recognition particle 14), resulting in a sixfold increase in cell-specific productivities for a low-yielding monoclonal antibody (mAb) [13]. Multitarget engineering approaches have been explored to address limitations in the secretory pathway of CHO cell lines. Cartwright et al. [24] overexpressed multiple ER proteins, identifying top-ranking effectors associated with increased titers and cell-specific productivities for an IgG1 (immunoglobulin G) LDH (lactate dehydrogenase) molecule [24]. For instance, Berger et al. [25] targeted the ER chaperone system, overexpressing PDI family members Erp27 and Erp57, leading to decreased accumulation of misfolded proteins, improved folding, and increased production of proteins such as IgG1, Fc-fusion protein, and interferon-β in CHO cells [25]. Another strategy is engineering the vehicle trafficking system to transport proteins from the ER to the cell surface. The production of secretory embryonic alkaline phosphatase was able to increase threefold when Peng et al. [26] overexpressed exocytic SNARE proteins such as SNAP-23 and VAMP8 [26]. Overexpressing ceramide transferase protein, involved in ER to Golgi transport, increased the output of tissue plasminogen activator, human serum albumin, and IgG in CHO cells [27].

3.2.4 Genetic engineering (knockdown or knockout strategies) of host cells for enhanced secretion

There are various activities of the discovered genes, and altering one gene may only have a little effect on the release of proteins. This highlights the need for multiple modifications, such as knockout, knockdown, or the coordinated overexpression of several genes [28]. Host cell engineering strategies have significantly improved productivity in biopharmaceutical production, involving traditional recombinant DNA technology and RNA interference for stable gene suppression. Recent advancements in omics technologies have provided a nuanced understanding of cellular physiology, leading to identify gene targets specific to host cell lines. Focus on the secretory machinery has revealed the importance of ER chaperones and proteins in enhancing antibody and recombinant protein productivity. The overexpression of specific chaperones, foldases, and global regulators like XBP-1s (X-box binding protein 1) has demonstrated effectiveness [29–31]. Fierro et al. (2023) presented a DiCre-based knock-sideways tool, knockER, enabling the conditional fusion of KDEL (Lys-Asp-Glu-Leu) to secreted proteins, redirecting them to the ER, and offering new possibilities for studying secreted proteins [32]. Additionally, introducing Blimp1β into CHO cells has improved the secretory capacity by activating vital elements of the unfolded protein response and mechanistic target of rapamycin (mTOR) signaling [33]. mTOR, an important protein involved in protein synthesis, is known to improve various cellular processes, including cell growth, proliferation, and survival. The specific productivity of therapeutic immunoglobulins is significantly increased by overexpression of mTOR in CHO cells [34]. While Zhang et al. [35] emphasize the importance of genetic engineering to alleviate ER stress and increase productivity [35], He et al. [36] discuss the emerging role of circular RNAs (circRNAs) in posttranscriptional and translational gene regulation, highlighting their interactions with RNA-binding proteins. Their research reveals that circRNAs can modulate gene expression, ultimately affecting mRNA translation [36].

Enhancing the secretory capacity of mammalian production cells through genetic engineering offers a promising approach to increase productivity. Studies have shown that overexpression of various genes involved in the secretory pathway or manipulation of cytoskeleton dynamics can significantly improve recombinant protein production [37]. These diverse approaches highlight the continuous innovation in host cell engineering to optimize recombinant protein production.

3.3 Engineering glycosylation

Glycosylation, a frequent posttranslational modification, affects various physiological and pathological processes, including migration, cell growth, tumor invasion, host–pathogen interaction, differentiation, transmembrane signaling, and cell trafficking. There is a substantial history of modifying glycosylation in various organisms, including mammalian cells, plants, fungi (yeast), and bacteria, using genetic techniques, resulting in a repertoire of well-characterized glycosylation mutants. The glycosylation engineering aims to modify therapeutic glycoproteins' size, charge, and solubility, preventing swift elimination from the bloodstream. Additionally, glycoengineering has enhanced the existing therapeutic approaches or created novel modalities. Glycans can act as ligands for lectin receptors, directing therapeutics to specific cells. Recent advancements, particularly in gene editing, offer precise control over glycosylation engineering, and the potential of the field appears boundless with expanding knowledge. The protein glycosylation can be divided into two main groups: *O*- and *N*-glycosylation. *O*-glycosylation is further divided into mucin-type and *O*-GlcNAcylation. The cotranslational and posttranslational processes involve glycosyltransferase enzymes, utilizing specific sugar nucleotide donors to attach carbohydrates to proteins, lipids, polypeptides, or polynucleotides [38–40].

Producing therapeutic enzymes for mammalian administration presents challenges due to specific glycosylation patterns crucial for ensuring correct biological activity, functionality, stability, and particularly low antigenicity. Over 50% of human proteins are recognized as potentially glycosylated, with many having multiple glycosylation sites. Protein glycosylation occurs at the carboxamide side chain of asparagines (Asns), glutamines (*N*-linked glycosylation), or the hydroxyl group of serine/threonine side chains (*O*-linked glycosylation) [41].

3.3.1 Types of glycosylation

3.3.1.1 *N*-linked glycosylation

N-glycosylation is a process initiated in the ER. Still, specific structural elements such as core fucosylation and branching are introduced later in the secretory pathway in the Golgi apparatus. *N*-glycans are modified by a series of sequentially acting glycosidases and glycosyltransferases that modify glycans in the Golgi apparatus and consequently dictate and ultimately determine the glycan profile of the whole cell. However, the structural modification of *N*-linked glycans is a complex process resulting in numerous glycan structures [42].

N-glycosylation plays a crucial role in the effector functions of IgG antibodies, and its clinical use includes therapeutic IgG antibodies with *N*-glycosylation tailored to enhance their cytotoxic properties [38]. *N*-Acetylglucosamine bound to asparagine (GlcNAcβ1-Asn) is the most common bond. The complexity and diversity of *N*-glycan structures and their implications in disease progression are still challenging. *N*-linked protein glycosylation begins with synthesizing the oligosaccharide precursor in the cytoplasm, which is then translocated to the ER lumen. The oligosaccharide precursors are transferred to the Asn residue of a newly formed protein when they undergo some modifications. Some trimming of the oligosaccharide chain is subsequently done in the ER, and the glycoprotein moves to the Golgi apparatus, where many additional sugar branches could be attached. *N*-glycosylation regulates the interaction of receptors and ligands with each other, coregulatory molecules, and distinct membrane domains in intact cells to alter signal transduction [38, 41].

Protein glycosylation encompasses the protein structure, secretory protein load, Golgi transport mechanism, enzyme protein levels, availability of monosaccharide nucleotides, and the organization of glycosylation enzymes within the Golgi apparatus [42].

3.3.1.2 *O*-linked glycosylation

O-glycosylation begins with the addition of *N*-acetylgalactosamine (GalNAc) to the hydroxyl group on certain serine or threonine residues. This process takes place in proteins traversing the Golgi compartment, facilitated by an enzyme known as *N*-acetyl galactosaminyltransferase (GalNAc-T), which transfers the GalNAc residue to the side chain of either serine or threonine. So far, it has been identified and characterized 15 unique members of the GalNAc-T family in mammals, potentially with up to 24 isozymes expressed in specific tissues. In *O*-linked protein glycosylation, posttranslational modifications start in the Golgi by attaching a GlcNAc molecule to the protein's serine or threonine residue. Next, multiple other sugars, such as sialic acid, can be added to the structure. *O*-glycosylation is crucial in various biological processes, such as pathogen interaction, cell adhesion, and proteolytic activities. Abnormal glycosylation has been linked to numerous diseases [43, 44].

O-glycosylation influences the protein structure and primarily affects surface-exposed regions. Mucin-type *O*-glycosylation, characterized by GalNac binding to serine/threonine residues, is predominantly found in mammalian-secreted and membrane-bound mucins. Bioinformatics tools such as NetOGlyc4.0 and ISOGlyP have been developed to estimate mucin-type *O*-glycosylation sites with approximately 75% and 70% accuracy, respectively [45, 46]. One of the primary challenges in *O*-glycobiology is still figuring out how the glycosyltransferases involved in *O*-glycan production in the secretory pathway are organized and vary dynamically [47].

3.3.2 Strategies for engineering glycosylation

Glycosylation is essential in many aspects of cancer development and progression, influencing tumor growth, evasion of immune surveillance, and metastasis, and also in determining the quality of recombinant glycoprotein therapeutics, particularly mAbs and protein stability [40, 48]. In cancer, glycosylation impacts critical processes like growth factor signaling, making it a target for biomarker identification. Glycosylation changes serve as indicators of stem-cell-like phenotypes within cancer and healthy tissues. Moreover, these alterations contribute to developing cancer-specific glycoproteins and can be exploited for therapeutic interventions. For instance, glycosylated biologics, such as otelixizumab and mogamulizumab, promise a therapeutic potential in treating conditions like T-cell lymphoma and type 1 diabetes, demonstrating the diverse applications of glycosylation research in cancer therapeutics [40].

Genetic approaches can alter cellular glycosylation abilities: knockdown and overexpression are the most commonly used genetic methods. Knockdown is not successful in mammalian cell lines as in plants and *Drosophila*. The most important reasons for this are low efficiency and glycosyltransferase activity. Overexpression introduces desirable glycosyltransferase activities through glycogene transfection. However, it lacks control over genomic integration sites, expression levels, and gene copy numbers, disrupting glycosylation patterns and instability of introduced traits. Precision genome editing tools like nucleases ZFN (zinc finger nuclease), transcription activator-like effector nucleases, and CRISPR (clustered regularly interspaced palindromic repeats)/Cas9 have overcome challenges in knockout and knock-in strategies, enabling specific and efficient gene manipulation across all cell types. These tools extend beyond gene knockout or knock-in to activate silent genes, mimic disease mutations, and insert foreign genes at precise genomic locations [49].

3.3.2.1 *N*-glycosylation engineering

In the realm of glycoengineering, knockout strategies, such as targeting the MGAT1 (alpha-1,3-mannosyl-glycoprotein 2-beta-*N*-c2cosaminyltransferase) gene, have been pivotal in modulating glycosylation patterns for enhanced antibody-dependent cellular cytotoxicity through increased binding affinity to the Fcγ-IIIa receptor [50]. Significant strides have been made in *N*-glycosylation pattern alterations, notably the elimination of core α1-6-fucosylation in CHO cells, achieved through overexpression of bisecting GlcNAcT-III (MGAT3) or gene editing to knockout FUT8 alleles. In human HEK293-T cells (immortalized human embryonic kidney cells), GlycoDelete produced antibodies with truncated *N*-glycans, affecting Fcγ receptor (FcγR) affinity. Efforts to achieve homogeneous α2-6-sialylation faced challenges, but the GlycoDelete strategy and precise gene editing in CHO cells provided a framework for enhancing *N*-glycan homogeneity [49]. The GlycoDelete strategy further reduced *N*-glycan structural heterogeneity, promoting galactosylation and sialylation and showcasing advancements in modifying glycosylation patterns with applications in therapeutic protein production [51]. Moreover, MGAT1 disruption in CHO cell lines, facilitated by ZFN-mediated genome editing, results in consistent Man-5 (*N*-linked mannose-5) glycan glycosylation, offering specificity and efficiency for producing therapeutic proteins targeting the mannose receptor in cancer vaccine applications [52]. Sialylated IgGs, bearing nine mutations in the MGAT1 gene, exhibit reduced FcγRIIIA affinity, shedding light on the glycosylation enzyme's structural intricacies and influencing antibody function [53]. These glycoengineering approaches collectively contribute to the precision and versatility of manipulating glycosylation for various therapeutic applications.

Glycosylation of enzyme levels critically shape glycan biosynthesis, influencing the characteristics of therapeutic proteins [42]. Overexpressing α2-6-sialyltransferases in CHO cells enhances antibody production with α2-6-linked sialic acids, mirroring human blood glycosylation patterns [54]. The GlycoDelete glycoengineering method streamlines mammalian cell Golgi *N*-glycosylation, yielding therapeutic proteins with simplified and sialylated trisaccharide *N*-glycan structures, offering improved consistency and performance [51]. Natural and artificial miRNAs were harnessed to regulate glycosylation in CHO cells, demonstrating their versatility for fine-tuning mAb fucosylation and influencing various glycan features [55]. Genetic engineering in mammalian cells significantly advances the understanding of genetic and biosynthetic control over the cellular glycome, highlighting the vital roles of glycans in development, health, and disease [56]. Various strategies such as small interfering RNA (siRNA) and small hairpin RNA have regulated sialidase gene expression in CHO cells, impacting interferon-gamma sialylation [57]. Glycoengineering techniques targeting the fucosyltransferase enzyme, such as knocking down or knocking out the FUT8 gene, enable the production of antibodies without fucosylation, presenting strategies for modifying protein *N*-glycosylation pathways in mammalian cells [58]. Mammalian cell lines like CHO, BHK (baby hamster kidney fibroblasts), NS0 (recombinant murine cells), and C127 (murine mammary epithelial tissue-derived cells) are extensively utilized in pharmaceutical protein production. However, efforts to engineer these cell lines aim at achieving consistent and specialized glycosylation patterns, emphasizing the importance of overexpressing glycosyltransferases for improved glycan structures in recombinant proteins [59].

3.3.2.2 *O*-glycosylation engineering

Despite various *O*-glycosylation types performed by mammalian cells, there is limited interest in *O*-glycans for recombinant therapeutics. Some clinically used coagulation factors and drugs, like erythropoietin (EPO) and Enbrel, carry *O*-GalNAc, *O*-Fuc (*O*-fucosylation), and *O*-Glc glycans. However, engineering *O*-GalNAc glycans is complex due to the involvement of up to 20 polypeptide GalNAc transferases.

In the realm of *O*-glycosylation engineering, researchers focus on manipulating CHO and HEK293 cells by knocking out and knocking in GALNT (polypeptide *N*-acetylgalactosaminyltransferases) genes, providing insights into the control of *O*-GalNAc glycans [49].

Concurrently, efforts in synthetic glycobiology and cell surface display systems involve the presentation of glycan pathways on cell surfaces for research on glycosylation biosynthesis genes and exploring biological functions associated with specific glycosylation structures [60]. Gene editing technologies, particularly CRISPR/Cas9, facilitate precise modifications in glycosylation-related genes, exemplifying targeted genome editing for refined glycoengineering [61]. These advancements collectively contribute to the diverse and precise landscape of glycoengineering technologies.

Prabhu et al. [50] utilized CRISPR/Cas9 to modulate glycosylation in CHO cells, achieving independent control over galactosylation and fucosylation in the expressed IgG. By knocking out enzymes related to uridine diphosphate galactose and guanosine diphosphate fucose synthesis in CHO cells expressing IgG, the researchers achieved predictable and independent control over galactosylation and fucosylation [50]. Wong et al. [62] investigated intracellular glycosylation activities in CHO cells, highlighting the cell line's historical significance in glycoengineering [62]. Reinl et al. [63] demonstrated Golgi engineering in CHO cells to express novel Asn-linked oligosaccharide structures. Knocking out essential glycosyltransferase genes (FUT8 and β4GALT1) and integrating synthetic glycosyltransferase genes into the genome allowed fine-tuning of fucosylation and galactosylation levels. The system regulated by small-molecule inducers provided simultaneous and independent control, influencing antibody binding to cell surface receptors and impacting effector functions. This engineered platform offers a versatile tool for tailored glycosylation [63].

Yang et al. [64] aimed to achieve glycan remodeling of recombinant EPO for homogeneous glycoforms. Despite attempts to modify EPO expressed from HEK293T cells using endoglycosidase, success was limited [64]. Konstantinidi et al. [39] utilized a glycoengineered cell-based platform to study O-glycosylation in mucin tandem repeat domains. They discovered that potential Ser/Thr O-glycosides were fully occupied when expressed in human embryonic kidney 293 SimpleCells, offering insights into mucin O-glycosylation. The methods established in this study allow for a more comprehensive examination of native mucins and shed light on the O-glycosylation of mucins and mucin-like domains [39]. Prati et al. [65] demonstrated that simultaneous regulation of enzyme activities in CHO cells represents a significant advance in metabolic engineering. They reported the suppressing of CHO cell ST3Gal1 (ST3 beta-galactoside alpha-2,3-sialyltransferase 1) gene expression and overexpression of human C2GnT (Core2 β-1,6-N-acetylglucosaminyltransferase) gene to regulate the O-glycosylation pathway [65]. Moreover, Dinter et al. [66] developed tricistronic vectors for stable expression of two glycosyltransferases in CHO cells, showcasing efficient glycosylation and biosynthetic efficiency. Thus, additional glycosylation routes can be provided to CHO cells by multicistronic vectors expressing glycosyltransferases, enabling the in vivo glycosylation of target proteins with the appropriate glycan moiety [66]. In addition, Heffner et al. (2021) discussed strategies to control O-glycosylation by altering GnT (N-acetylglucosaminyltransferase) activity. Research indicated that elevating C2GnT levels in CHO DG44 cells might impact T-cell activation. Additionally, combining C2GnT overexpression with ST3Gal1 suppression could alter the O-glycosylation pathways. This study implies that cellular engineering can concurrently adjust the levels of competing glycosylation enzymes [67]. Collectively, these studies contribute to developing diverse and precise glycoengineering approaches with implications for therapeutic protein development.

3.4 Engineering cellular communication

Cellular communication has a significant role in the life cycle of a cell. It affects the signaling pathways that control cellular differentiation, proliferation, metabolism, and apoptosis. As cellular communication affects various pathways and causes numerous different responses, it can be controlled in many ways, resulting in a wide range of options for cancer treatment [68].

3.4.1 Extracellular vesicles

EVs are crucial for cancerous cellular communication, considering they provide signaling via packaging and delivery of bioactive constituents from one cell to another as well as they are an important target to optimize protein secretion as mentioned above [69]. The secretory pathway, a crucial facet of the vesicle trafficking system, provides the precise and regulated distribution of proteins, metabolites, and molecular cargo among and across cellular compartments [70]. Therefore, exosomes can be manipulated via engineering secretion frequency, packaging specificity, cytosolic merge ability, and targeting of exosomes to be used for natural carriers of various molecules [71]. For instance, Kojima et al. [72] created synthetic-biology-inspired EXOtic (EXOsomal transfer into cells) devices (Figure 3.2) that are created by manipulating all these aspects in a time of need. It consists of a system creating EVs that package expressed mRNA to release for targeting cells, producing packaged mRNA delivered and internalized through exosomes [69, 72]. Similarly, smart exosomes were developed [73] to target tumor-draining lymph nodes by engineering the exosome surface to exhibit L-selectin and OX40 ligand [73]. It is a comprehensive study since the cytosolic merge ability and targeting were achieved at the same time.

Moreover, immature dendritic cells (DCs) of mouse were engineered to generate exosomes with αv integrin-specific iRGD peptide (CRGDKGPDC) on the surface by genetically fusing the peptide and exosomal membrane protein (Lamp2b) for enhanced tumor targeting, and the loaded doxorubicin was delivered to the tumor site with enhanced specificity [74]. Therefore, EVs can be a promising tool to tune cellular communication as a more specific cancer therapy option.

3.4.2 Manipulating direct cell contact for diagnosis and therapy

Mammalian cells can communicate directly with molecules on the surface of other cells through proteins, fats, and carbohydrates, and their various complexes were displayed on their cell membranes. This causes similar or different signaling pathways to be affected in both cells and cellular responses [75]. Chimeric receptors can be used to stimulate directed cell migration, which can be a promising approach for reduce-

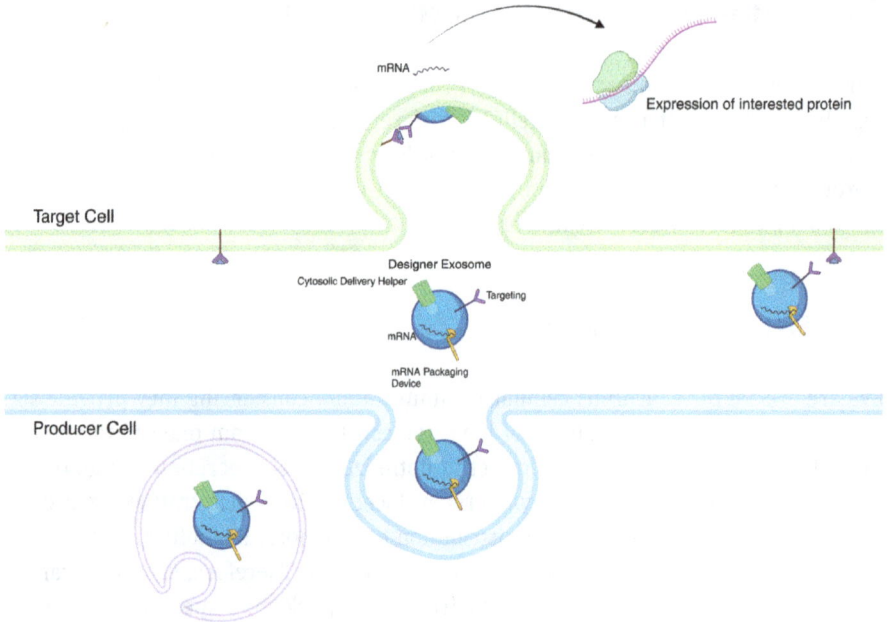

Figure 3.2: Schematic representation of the exotic device. It exhibits cytosolic delivery helper, targeting protein, and mRNA packaging devices on the surface of the designer exosome that packaged mRNA. Target cells internalize exosomes after recognition and cell membrane assembly of the exosome, and mRNA is translated to express the target protein in the target cell [69, 72].

ment of proximity between the therapeutic and the target cell. For example, fused IL-6R extracellular domain to vascular endothelial growth factor receptor intracellular domain on cells can be used for targeting IL-6-secreting senescent cells as a result of activated ros (Ras homolog family member A) GTPase activation [76]. Most of the direct cell contact-based therapies rely on the engineered cell therapies.

Mesenchymal stem cells (MSCs) are nonhematopoietic multipotent cells, which have the potential to differentiate into chondrogenic, osteogenic, and adipogenic cells. It is known that these cells display some specific antigens on their membranes and can migrate to tumor areas for the purpose of tissue repair and reconstruction. Therefore, in treatments where immune surveillance is maintained, the possible use of MSCs in targeting tumors and delivering potential cargo has brought to the fore their potential for treatment and diagnosis. For example, in thyroid cancer, MSCs engineered to express sodium iodide symporter (NIS) can be used to target iodide-concentrating thyroid glands, and this system can be used for imaging purposes with radioiodine labeling and also in the treatment of tumors in combination with radionuclide. There are even studies showing that imaging in other types of cancer can be done successfully in this way by providing ectopic NIS expression. A limiting factor for this relatively targeted system may be the damage experienced by off-target cells following MSC recruitment. To avoid this situation, Schug et al. [78] developed a system that allows NIS expression to be selectively induced and is sensitive to TGFB1-inducible SMAD under hypoxia conditions, thus planning to increase the effect of targeting [77, 78]. Currently, Mayo Clinic conducted a treatment for patients with recurrent ovarian cancer in phase II as MSCs expressing NIS were infected with a genetically engineered measles virus [79]. Similarly, MSC-TRAIL (MSC-tumor necrosis factor-related apoptosis-inducing ligand) presents a targeted treatment of lung cancer through engineered MSCs to express TRAIL, which specifically promotes apoptosis in cancer cells in phase II [80].

DCs are immune cells that secrete antigens on their cell surfaces, stimulating the immune system in the presence of a possible pathogen or cancer cell, thereby creating an immune response. In addition, DCs have been studied for decades as an active immunotherapy in cancer treatments due to their ability to promote T-cell activation, apart from other antigen-presenting cells, and to be transported to lymph nodes, even if limited [81, 82]. In this context, Koch et al. [84] created a system including DCs transfected with IKKβ-activated RNA for T-cell and Natural killer (NK) cell activation to treat uveal melanoma, and this vaccine is in phase I trial [83, 84]. DC implantation can also be used as a supportive treatment. For example, it has been observed that when DCs are injected before nephrectomy in patients with metastatic renal carcinoma, they boost the immune response by inducing dense infiltration of T cells and may therefore increase the survival [81]. In addition, DCs are being investigated for use in the treatment of brain cancer because they can cross the barriers of the central nervous system. On this level, a phase III study supported by Oslo University Hospital to treat glioblastoma, which has a very high mortality rate, involves glioblastoma stem cells, mRNA transfected with DC vaccines injected together with the antiapoptotic peptide survivin, and human telomerase reverse transcriptase [85].

CAR-T (chimeric antigen receptor (CAR) T cells) are the most advanced immunotherapeutic treatment method among all chimeric antigen receptor-based treatments. Essentially, it involves activating T cells taken from the patient and reengineering them through CAR. Kymriah, Yescarta, and Abecma are among some CAR-T therapies approved by the FDA. Genetic engineering of CAR-T cells must be done very stringently to avoid toxicity in the body. One of the existing methods to do this is to control CAR expression through the inclusion of ON and OFF switches and molecules such as tetracycline and doxycycline. In addition, drug-inducible caspase-9 (iCasp9) is used as a kill switch and induces the apoptosis pathway. In addition, antigen escape and tumor antigen heterogeneity can be prevented by methods using improved circuits such as AND, OR, and NOT gates, which are used to increase the effectiveness and specificity of tumor targeting. Another approach involves genetic arrangements that enable CAR-T cells to interfere with immunologic factors in target cells. In fact, T cells obtained with this method are also known as Armored CAR-T cells. An example is an approach in which immune-checkpoint inhibitory molecules in cancer cells are inhibited by CAR-T cells via CRISPR-Cas9 or siRNA. Although the effectiveness of CAR-T has been proven, it has some limitations due to the costs and disruptions that may occur during the processing of T cells taken from the patient. In addition, there is a need for more tuned and controllable systems to prevent tumor antigen escape. Interaction of CAR-T with normal and cancer cells can be intervened by genetic engineering through tuning of CAR expression and activation profile (Figure 3.3) [1, 86].

3.5 Engineering cellular metabolism

Mammalian cell metabolism encompasses producing and using energy necessary for cell growth, functionality, and survival. These processes involve transforming nutrients into energy and building blocks by cells through various chemical reactions [90]. The basic steps include glycolysis, pyruvate oxidation, tricarboxylic acid (TCA) cycle (Krebs cycle), oxidative phosphorylation, lipolysis, beta-oxidation, amino acid catabolism, pentose phosphate pathway, gluconeogenesis, and glycogenolysis [91]. Glycolysis is responsible for the production of energy and pyruvate, a valuable intermediate molecule transported to the mitochondria to be used in steps such as the Krebs cycle. Glycolysis is essential because it causes rapid energy provision and has a place in cancer metabolism [92]. During the oxidation of pyruvate followed by glycolysis, pyruvate is converted to acetyl-CoA, and carbon dioxide is released. Another fundamental step, the TCA cycle, plays a central role in energy production in the presence of oxygen. In the TCA cycle, NADH (reduced nicotinamide adenine dinucleotide) and FADH2 (flavin adenine dinucleotide) molecules are formed to be utilized for energy production during oxidative phosphorylation. In addition, the output of many intermediate products required for other metabolic pathways, such as gluconeogenesis, lipogenesis, and amino acid synthesis, is also achieved at this stage. Oxidative phosphorylation involves the oxidation of NADH and FADH2 molecules obtained in previous reactions in the cell during the electron transport chain (ETC) and the conversion of the proton gradient resulting from electron flow into ATP (adenosine triphosphate) by the ATP synthase enzyme. Water is released using oxygen during this highly energy-efficient phase [93]. At lipolysis and beta-oxidation, acetyl-CoA is produced as a result of the oxidation of fatty acids during fat lysis and is converted to ketogenic intermediates [94]. Amino acid catabolism is also a critical step, where carbon skeletons and ammonia are formed. In addition to contributing to energy production by entering the TCA cycle, carbon skeletons can also be used in protein synthesis [95]. The pentose phosphate pathway works parallel with glycolysis, producing ribose-5-phosphate necessary for nicotinamide adenine dinucleotide phosphate and nucleic acid synthesis. Finally, gluconeogenesis enables glucose production from various sources and glycogenolysis from glycogen, allowing cell metabolism to produce energy when required. When viewed this way, many intermediate molecules can be created in more than one stage and used in various stages, revealing the complexity of mammalian cell metabolism [96]. Mammalian cellular metabolism is an integrated

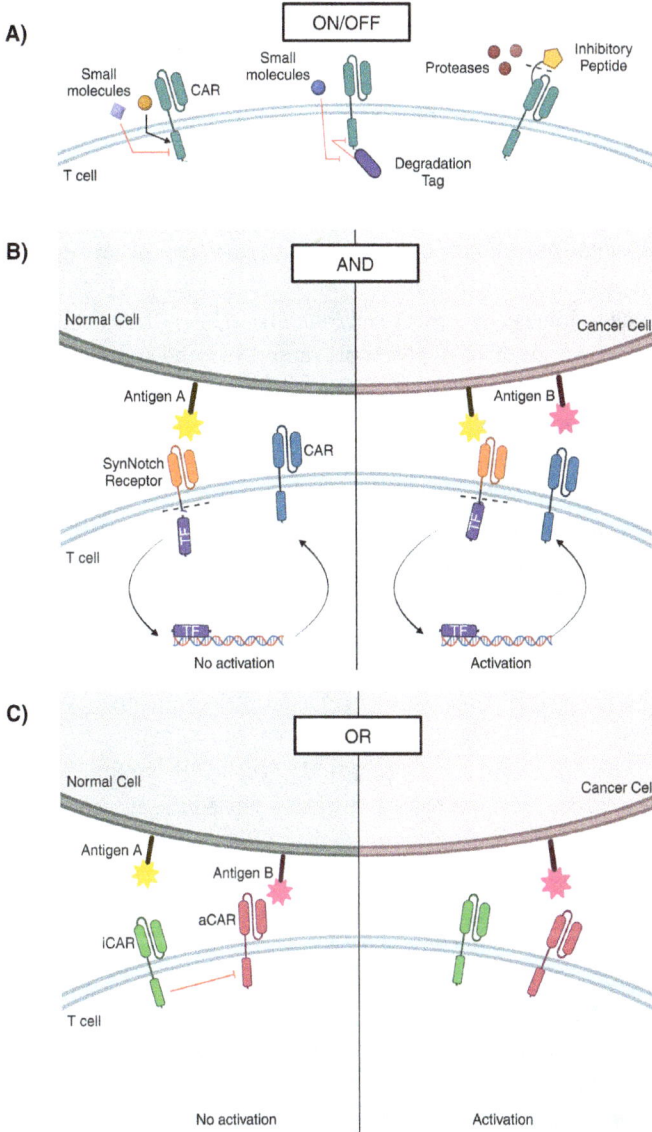

Figure 3.3: Schematic representation of CAR-T cell engineering methods to increase efficiency and specificity. (A) ON/OFF switches for controlling the expression and activity of CARs. Small molecules can be used for controlling the expression of CARs on the surface. Also, small molecules or low levels of O_2 can inhibit the degradation tags that CARs fused to activate the system with control, which is initially found in the OFF position. In addition, inhibitory peptides can be fused to the CARs where proteases are found in the tumor microenvironment to activate them on the tumor site. (B) An example of the use of AND gates in CAR-T technology. Upon recognition of antigen A with scFv extracellular domain, transcription factor (TF) on the synNotch receptor is cleaved and activates the transcription of CAR, yet this system only gets activated when antigen B is present since recognition of it with CAR is necessary for T-cell activation. (C)

system that ensures efficient energy production and use. Cells can regulate and adapt these pathways according to energy needs [97].

Cancer metabolism has become a studied and known phenomenon since Otto Warburg's breakthrough in the early twentieth century. The model developed by Warburg based on his observations on cancer cells was called the Warburg effect or aerobic glycolysis and was based on the fact that cancer cells, unlike normal cells, use glycolysis even in the presence of oxygen [98]. At that time, it was thought that cancer cells preferred glycolysis over mitochondrial respiration. It has been observed that oncogenes and pathways such as MYC, AKT, and mTOR increase glycolysis in the cancer scenario [99]. Also, due to hypoxia in cancer cells, the formation of ATP from substrates is directed to be carried out through glycolysis rather than the TCA cycle. The increased tendency to glycolysis is associated with negative feedback from accumulating irreducible NADH. However, this situation, which occurs due to inhibition of the TCA cycle due to hypoxia, should not be confused with NADH accumulation due to ETC error, as this may result in growth impairment and cell death [100]. In line with this, it is known today that mitochondrial respiration and the TCA cycle are still necessary for producing metabolites, such as oxaloacetate, which are essential for supporting tumor development [101]. This requirement is reinforced by the observation that nucleotide synthesis is also supported by glycolysis, one-carbon metabolism, and the TCA cycle in recent studies in pancreas and lung cancers with KRAS (Kirsten rat sarcoma virus gene) and p53 loss profiles [102]. This whole order is accompanied by reactive oxygen species (ROS), and cancer growth is promoted due to oxidative stress altering the metabolism [91]. Since cancer cells constantly develop, they require antioxidant supplements to increase survival through intensive nucleotide synthesis and removing toxic products [95]. Additionally, it is known that the cell exhibits a different metabolic profile during metastasis than that in a benign mass. This situation can be explained by the fact that the location of primary cancer cells, which try to dominate healthy cells in the secondary tumor area, is shaped by the metabolism according to the tissues' nutritional profile and the primary cells' capacity to adapt to this profile.

It is known that in some types of cancer, high expression of genes that interfere with the metabolic flow, such as MTC1, and genes that promote epithelial–mesenchymal transition and high production of ROS metabolites contribute to this profile [103]. In addition, the metabolic processes of cancer cells trying to adapt to the secondary tumor site can be shaped according to the nutritional content of this area. For example, when breast cancer cells invade the brain, their phosphoglycerate dehydrogenase expression increases, which enables their brain cells to make a metabolic adaptation regarding

Figure 3.3 (continued)

An example of the use of OR gates in CAR-T technology. Upon recognition of antigen A with iCar (inhibitory CAR), iCAR inhibits the aCAR (activating CAR), and this results in inactivation yet in the presence of antigen B only, iCAR cannot suppress the aCAR, and T-cell is activated. These two gate usages in CAR-T cells represent the tuning of CAR-T specificity toward cancer cells [87–89].

nutritional intake and nutrients in the environment, which increases their survival. Identifying targets that increase survival by causing such metabolic alteration creates targets for drugs that can be developed against metastatic cancers [104].

A good understanding of cancer metabolism is important for developing cancer cell lines and media optimization of these cell lines. For example, Akt or cMyc over-expression is found to be related to higher glucose uptake and lactate production. Therefore, manipulation of Akt or cMyc activations in cell lines such as CHO and NS0 can prevent lactate production, which burdens the cell, or alter the glucose flux [105]. Also, BHK and HEK293 cells were engineered to express cytosolic pyruvate carboxylase, resulting in increased viability and production. Moreover, partial knockout of LDH in hybridoma cells was found to be related to an increased titer and reduced lactate production [106].

Engineering metabolism may also have certain benefits in cell-based therapies. This may increase cell survival or effectiveness. For example, it has been observed that reprogramming the metabolism of CAR-T and CAR-NK cells with stimulants such as CD28 increases the viability of CAR-T. Similarly, knockdown of CAR-IL-15 and cytokine-inducible SH2-containing protein in NK cells was observed to promote the glycolytic pathway and lead to higher in vivo activity [107]. It is known that even anti-inflammatory cytokine activity can be inhibited with NK cells via alteration in their metabolism [108].

3.6 Conclusion

Glycosylation is crucial in cancer development and progression and determines the quality of recombinant glycoprotein therapeutics. The ability of cellular glycosylation can be modified and improved by genetic methods. Unique synthetic biology methods enable specific and effective gene manipulations. Glycosylation is often viewed as posttranslational modification. The methods focusing on different aspects are examined further. First, N-glycosylation engineering focuses on increasing the production of therapeutic proteins through methods such as targeting the MGAT1 gene and the GlycoDelete method. Second, although O-glycosylation engineering has generally received limited interest for recombinant therapeutics, some clinically useful coagulation factors and drugs carry O-GalNAc, O-Fuc, and O-Glc glycans. Glycosylation engineering has come to the fore in tailoring glycan structures to achieve optimal therapeutic protein properties. Gene editing techniques such as CRISPR/Cas9, ZFNs, and siRNAs help to holistically unravel the molecular mechanisms governing glycosylation. Engineering protein release and glycosylation patterns in mammalian cells have gained significant importance, especially in therapeutics.

The secretion of proteins requires several steps to proceed within the ER and Golgi apparatus. In this process, carrier vesicles pass from the donor compartment to the target compartment, and in this transition, the interaction of COPI, COPII, clathrin, SNARE, and Sec1/Munc18 proteins is required for vesicle fusion. In this whole process, all proteins and pathways that play an active role can be directed to the desired purpose using synthetic biology techniques. In addition, secretion mechanisms can also be engineered directly. Optimization of signal peptides in CHO cells can increase the secretion of recombinant proteins, and this process involves optimization of signal peptides to increase the production efficiency of recombinant biotherapeutic antibodies. Manipulating vesicle transport plays a critical role in maintaining organelle homeostasis in eukaryotic cells and can increase protein secretion. These studies are frequently encountered in the literature as a result of the potential for knockdown, knockout, or overexpression of identified genes to significantly increase recombinant protein production. Thanks to these advances and genetic engineering methods that increase the production of recombinant proteins, optimize the secretion profiles of the produced proteins, and enable the developed proteins to emerge as high-quality products; cellular mechanisms have been better understood; and significant progress has been made to increase the efficiency.

Cellular communication is crucial for controlling differentiation, proliferation, and apoptosis. This complex communication system provides many options for possible cancer therapies. EVs, crucial for communication between cells, can be modified to transport active substances, offering precise cancer treatments. New creations such as synthetic biology-based EXOtic devices and intelligent exosomes demonstrate the possibilities of controlling vesicle trafficking for medical uses. Yet, there is a need for further development in the utilization of EVs since the actual concentration of vesicles that can reach the target points is generally low. Furthermore, the use of direct cell contact and chimeric receptors shows the potential for boosting therapeutic cell movement and performance in tumor settings. The difficulty of engineering cell-to-cell contact engages with the variations between patients, which creates a need for personalized medicine. Also, it rises the necessity for a precise tuning since any efficient application would require more than one circuit. Still, these developments highlight the potential of cell communication engineering in creating advanced and efficient cancer therapies in the future. Engineered cell-based therapies hold great promise. Yet, there are certain advantages and disadvantages of different cell types. MSCs can be off-target, and this may lead to severe side effects or misleading scanning results. The ability of MSCs toward targeting should be increased with tumor-specific engineering, and this requires an understanding regarding MSC tumor tropism. DCs are also important in terms of immunotherapy, especially for brain cancer types, since they can pass the barriers, yet their representation of predetermined antigens on the surface may lead to decreased efficiency in terms of recognition ability for further mutated cells. Yet, this can be turned into an advantage with the future advancements on sequencing and personalized medicine since low mutations will lead to higher responses on the individual this time. Also, restrictions of DCs to migrate to the lymph system might require conditioning of the cancer area to promote migration as a previous study [109] demonstrated that preapplication of tetanus/diphtheria toxoid promotes DC migration in the cancer area [109]. Similarly, survival of glioblastoma patient was increased with a recall antigen injection including enhanced DC migration [81]. Therefore, DC vaccination stands as a promising way to treat early stage cancer patients despite these limitations. Cell-based therapies stand out due to their potential to enable safe and effective therapy beyond other methods, thanks to redirected cell functions through methods such as gene therapy. Gene therapy has some translational limits due to problems such as integration, localization, and frequency that occur during transgene delivery. Techniques developed to prevent this – for example, controlled transgene integration with engineered transposase – will contribute to eliminating the difficulties of cell-based cancer treatments.

Treatment techniques can be customized to disrupt specific metabolic pathways necessary for cancer cell survival and proliferation by identifying metabolic route dependencies in cancer cell lines and describing metabolic subtypes [110]. However, this poses some difficulties. First, this strategy may become particularly challenging as key elements of the immune system, such as macrophages and B cells, also undergo metabolic reprogramming with antigen recognition and stimulation. As a specific example of this situation, it has been observed that T cells' tendency to glycolysis increases as a result of antigen recognition and stimulation. Although this is promising for the treatment of autoimmune diseases, it may reduce the immune response and may cause metabolic engineering-related methods not to be preferred in cancer treatment [111]. In this area, changes in diet, in addition to treatments, have also shown potential in halting cancer progression. For example, high-fat diets may contribute to the development of fatty acid-preferring metastatic cells, which may complicate the patient's treatment. This situation, combined with the patient's genetic characteristics, may make treatment difficult for cancer metabolism. Therefore, advances in cancer metabolism are needed to develop a personalized medicine approach. In this way, methods that can reduce side effects in cancer treatment and can actually be quite effective are identified. In addition, the relationship between dormancy and metabolites, which can sometimes be seen in the case of metastasis, requires further investigation [112]. On the other hand, engineering cell metabolism can be a complicated task because healthy cells can also be quite sensitive to the targeted treatment approach. However, development in this field is used to increase the product yield and cell mass.

Engineering mammalian cells presents several challenges despite the advancements in genetic engineering technologies. This process is hindered by inexplicable failures, limiting the efficiency of circuit engineering in these cells. Additionally, the complexity of mammalian cell systems, including limited experimental data and complex regulatory mechanisms, presents obstacles in metabolic network analysis and engineering [113]. Furthermore, intercellular variation impacts therapeutic efficacy and safety, influencing allogeneic cell-based therapy development caused by risks like graft-versus-host disease. To overcome these difficulties, engineering mammalian cells for cancer needs to prioritize the versatility and adaptability of them while considering potential trade-offs in therapeutic applications. For example, utilizing prokaryotic two-component regulatory systems as tools for creating orthogonal signaling pathways in mammalian cells shows the dose-dependent transduction of small-molecule ligands [114]. Utilization of mutually orthogonal aminoacyl-tRNA synthetase in mammalian cells demonstrated precise control over protein engineering [115]. These systems have prioritized enabling specific communication between engineered cells for various applications, but the trade-offs associated with insufficient orthogonal signaling pathways emphasized the limited implementation of these tools for therapeutic applications. In summary, trade-offs between engineering tools based on current advancement and unsolved potential challenges require crucial considerations and further prioritizations of some qualities in the design of mammalian cells for cancer treatment. The application of engineering tools based on synthetic biology promises potential solutions for current trade-offs and setbacks [116].

Overall, the conventional method of focusing on individual genes in cell manipulation is changing, as there is a growing understanding of the necessity to manipulate multiple genes in different cellular pathways. Understanding the interconnected cellular processes in mammalian cells helps to tackle the complexity of engineering mammalian cells. Yet, this complexity can also provide various alternatives for new approaches.

References

[1]   Zhao N, Song Y, Xie X, Zhu Z, Duan C, Nong C, et al. Synthetic biology-inspired cell engineering in diagnosis, treatment, and drug development. Signal Transduct Target Ther 2023;8:1–21. https://doi. org/10.1038/s41392-023-01375-x.

[2]   Tasdogan A, Faubert B, Ramesh V, Ubellacker JM, Shen B, Solmonson A, et al. Metabolic heterogeneity confers differences in melanoma metastatic potential. Nature 2020;577:115–20. https://doi.org/10.1038/s41586-019-1847-2.

[3]   Cavenee WK, White RL. The genetic basis of cancer. Sci Am 1995;272:72–79. https://doi.org/10.1038/ scientificamerican0395-72.

[4]   Bray F, Laversanne M, Sung H, Ferlay J, Siegel RL, Soerjomataram I, et al. Global cancer statistics 2022: GLOBOCAN estimates of incidence and mortality worldwide for 36 cancers in 185 countries. CA Cancer J Clin 2024;74:229–63. https://doi.org/10.3322/caac.21834.

[5]   Bolli N, Payne EM, Rhodes J, Gjini E, Johnston AB, Guo F, et al. cpsf1 is required for definitive HSC survival in zebrafish. Blood 2011;117:3996–4007. https://doi.org/10.1182/blood-2010-08-304030.

[6]   Miller JD, Stacy T, Liu PP, Speck NA. Core-binding factor β (CBFβ), but not CBFβ–smooth muscle myosin heavy chain, rescues definitive hematopoiesis in CBFβ-deficient embryonic stem cells. Blood 2001;97:2248–56. https://doi.org/10.1182/blood.V97.8.2248.

[7]   Ye H, Fussenegger M. Synthetic therapeutic gene circuits in mammalian cells. FEBS Lett 2014;588:2537–44. https://doi.org/10.1016/j.febslet.2014.05.003.

[8]   Encyclopedia of Biological Chemistry, Academic Press; 2013.

[9]   Lakkaraju AKK, Thankappan R, Mary C, Garrison JL, Taunton J, Strub K. Efficient secretion of small proteins in mammalian cells relies on Sec62-dependent posttranslational translocation. Mol Biol Cell 2012;23:2712–22. https://doi.org/10.1091/mbc.e12-03-0228.

[10] Gutierrez J, Feizi A, Li S, Kallehauge T, Hefzi H, Grav L, et al. Genome-scale reconstructions of the mammalian secretory pathway predict metabolic costs and limitations of protein secretion. Nat Commun 2020;11:68. https://doi.org/10.1038/s41467-019-13867-y.

[11] Peng R-W, Fussenegger M. Molecular engineering of exocytic vesicle traffic enhances the productivity of Chinese hamster ovary cells. Biotechnol Bioeng 2009;102:1170–81. https://doi.org/10.1002/bit.22141.

[12] Owji H, Nezafat N, Negahdaripour M, Hajiebrahimi A, Ghasemi Y. A comprehensive review of signal peptides: Structure, roles, and applications. Eur J Cell Biol 2018;97:422–41. https://doi.org/10.1016/j.ejcb.2018.06.003.

[13] Le Fourn V, Girod P-A, Buceta M, Regamey A, Mermod N. CHO cell engineering to prevent polypeptide aggregation and improve therapeutic protein secretion. Metab Eng 2014;21:91–102. https://doi.org/10.1016/j.ymben.2012.12.003.

[14] Kober L, Zehe C, Bode J. Optimized signal peptides for the development of high expressing CHO cell lines. Biotechnol Bioeng 2013;110:1164–73. https://doi.org/10.1002/bit.24776.

[15] Park J-H, Lee H-M, Jin E-J, Lee E-J, Kang Y-J, Kim S, et al. Development of an in vitro screening system for synthetic signal peptide in mammalian cell-based protein production. Appl Microbiol Biotechnol 2022;106:3571–82. https://doi.org/10.1007/s00253-022-11955-6.

[16] Bachhav B, de Rossi J, Llanos CD, Segatori L. Cell factory engineering: Challenges and opportunities for synthetic biology applications. Biotechnol Bioeng 2023;120:2441–59. https://doi.org/10.1002/bit.28365.

[17] Cheng K-W, Wang F, Lopez GA, Singamsetty S, Wood J, Dickson PI, et al. Evaluation of artificial signal peptides for secretion of two lysosomal enzymes in CHO cells. Biochem J 2021;478:2309–19. https://doi.org/10.1042/BCJ20210015.

[18] Zhao L, Poschmann G, Waldera-Lupa D, Rafiee N, Kollmann M, Stühler K. OutCyte: A novel tool for predicting unconventional protein secretion. Sci Rep 2019;9:19448. https://doi.org/10.1038/s41598-019-55351-z.

[19] Vlahos AE, Call CC, Kadaba SE, Guo S, Gao XJ. Compact programmable control of protein secretion in mammalian cells. bioRxiv 2023;2023, 10.04.560774. https://doi.org/10.1101/2023.10.04.560774.

[20] Mansouri M, Ray PG, Franko N, Xue S, Fussenegger M. Design of programmable post-translational switch control platform for on-demand protein secretion in mammalian cells. Nucleic Acids Res 2023;51:e1. https://doi.org/10.1093/nar/gkac916.

[21] Praznik A, Fink T, Franko N, Lonzarić J, Benčina M, Jerala N, et al. Regulation of protein secretion through chemical regulation of endoplasmic reticulum retention signal cleavage. Nat Commun 2022;13:1323. https://doi.org/10.1038/s41467-022-28971-9.

[22] Cui L, Li H, Xi Y, Hu Q, Liu H, Fan J, et al. Vesicle trafficking and vesicle fusion: Mechanisms, biological functions, and their implications for potential disease therapy. Mol Biomed 2022;3:29. https://doi.org/10.1186/s43556-022-00090-3.

[23] Meldolesi J. Unconventional protein secretion dependent on two extracellular vesicles: Exosomes and ectosomes. Front Cell Dev Biol 2022;10:877344. https://doi.org/10.3389/fcell.2022.877344.

[24] Cartwright JF, Arnall CL, Patel YD, Barber NOW, Lovelady CS, Rosignoli G, et al. A platform for context-specific genetic engineering of recombinant protein production by CHO cells. J Biotechnol 2020;312:11–22. https://doi.org/10.1016/j.jbiotec.2020.02.012.

[25] Berger A, Le Fourn V, Masternak J, Regamey A, Bodenmann I, Girod P-A, et al. Overexpression of transcription factor Foxa1 and target genes remediate therapeutic protein production bottlenecks in Chinese hamster ovary cells. Biotechnol Bioeng 2020;117:1101–16. https://doi.org/10.1002/bit.27274.

[26] Peng R-W, Abellan E, Fussenegger M. Differential effect of exocytic SNAREs on the production of recombinant proteins in mammalian cells. Biotechnol Bioeng 2011;108:611–20. https://doi.org/10.1002/bit.22986.

[27] Torres M, Hussain H, Dickson AJ. The secretory pathway – The key for unlocking the potential of Chinese hamster ovary cell factories for manufacturing therapeutic proteins. Crit Rev Biotechnol 2023;43:628–45. https://doi.org/10.1080/07388551.2022.2047004.

[28] Idiris A, Tohda H, Kumagai H, Takegawa K. Engineering of protein secretion in yeast: Strategies and impact on protein production. Appl Microbiol Biotechnol 2010;86:403–17. https://doi.org/10.1007/s00253-010-2447-0.

[29] Lim Y, Wong NSC, Lee YY, Ku SCY, Wong DCF, Yap MGS. Engineering mammalian cells in bioprocessing – Current achievements and future perspectives. Biotechnol Appl Biochem 2010;55:175–89. https://doi.org/10.1042/BA20090363.

[30] Latorre Y, Torres M, Vergara M, Berrios J, Sampayo MM, Gödecke N, et al. Engineering of Chinese hamster ovary cells for co-overexpressing MYC and XBP1s increased cell proliferation and recombinant EPO production. Sci Rep 2023;13:1482. https://doi.org/10.1038/s41598-023-28622-z.

[31] Wang X, Kang L, Kong D, Wu X, Zhou Y, Yu G, et al. A programmable protease-based protein secretion platform for therapeutic applications. Nat Chem Biol 2024;20:432–42. https://doi.org/10.1038/s41589-023-01433-z.

[32] Fierro, M. A., Hussain, T., Campin, L. J., & Beck, J. R. (2023). Knock-sideways by inducible ER retrieval enables a unique approach for studying Plasmodium-secreted proteins. Proceedings of the National Academy of Sciences, 120(33), e2308676120.

[33] Kim SH, Baek M, Park S, Shin S, Lee JS, Lee GM. Improving the secretory capacity of CHO producer cells: The effect of controlled Blimp1 expression, a master transcription factor for plasma cells. Metab Eng 2022;69:73–86. https://doi.org/10.1016/j.ymben.2021.11.001.

[34] Dreesen IAJ, Fussenegger M. Ectopic expression of human mTOR increases viability, robustness, cell size, proliferation, and antibody production of Chinese hamster ovary cells. Biotechnol Bioeng 2011;108:853–66. https://doi.org/10.1002/bit.22990.

[35] Zhang L, Gao J, Zhang X, Wang X, Wang T, Zhang J. Current strategies for the development of high-yield HEK293 cell lines. Biochem Eng J 2024;205:109279. https://doi.org/10.1016/j.bej.2024.109279.

[36] He AT, Liu J, Li F, Yang BB. Targeting circular RNAs as a therapeutic approach: Current strategies and challenges. Signal Transduct Target Ther 2021;6:1–14. https://doi.org/10.1038/s41392-021-00569-5.

[37] Fischer S, Otte K. CHO Cell Engineering for Improved Process Performance and Product Quality. In: Lee GM, Faustrup Kildegaard H, Lee SY, Nielsen J, Stephanopoulos G, editors. Cell Cult. Eng. 1st ed., Wiley; 2019, pp. 207–50. https://doi.org/10.1002/9783527811410.ch9.

[38] Loaeza-Reyes KJ, Zenteno E, Moreno-Rodríguez A, Torres-Rosas R, Argueta-Figueroa L, Salinas-Marín R, et al. An overview of glycosylation and its impact on cardiovascular health and disease. Front Mol Biosci 2021;8. https://doi.org/10.3389/fmolb.2021.751637.

[39] Konstantinidi A, Nason R, Čaval T, Sun L, Sørensen DM, Furukawa S, et al. Exploring the glycosylation of mucins by use of O-glycodomain reporters recombinantly expressed in glycoengineered HEK293 cells. J Biol Chem 2022;298. https://doi.org/10.1016/j.jbc.2022.101784.

[40] Reily C, Stewart TJ, Renfrow MB, Novak J. Glycosylation in health and disease. Nat Rev Nephrol 2019;15:346–66. https://doi.org/10.1038/s41581-019-0129-4.

[41] Gómez S, Fernández FJ, Vega MC. Heterologous Expression of Proteins in *Aspergillus*. In: Gupta VK, editor. New Future Dev. Microb. Biotechnol. Bioeng., Amsterdam: Elsevier; 2016, pp. 55–68. https://doi.org/10.1016/B978-0-444-63505-1.00004-X.

[42] Frontiers | The N-Glycosylation Processing Potential of the Mammalian Golgi Apparatus n.d. https://www.frontiersin.org/articles/10.3389/fcell.2019.00157/full (accessed June 2, 2024).

[43] Gómez S, Fernández FJ, Vega MC. Heterologous Expression of Proteins in *Aspergillus*. In: Gupta VK, editor. New Future Dev. Microb. Biotechnol. Bioeng., Amsterdam: Elsevier; 2016, pp. 55–68. https://doi.org/10.1016/B978-0-444-63505-1.00004-X.

[44] Loaeza-Reyes KJ, Zenteno E, Moreno-Rodríguez A, Torres-Rosas R, Argueta-Figueroa L, Salinas-Marín R, et al. An overview of glycosylation and its impact on cardiovascular health and disease. Front Mol Biosci 2021;8. https://doi.org/10.3389/fmolb.2021.751637.

[45] Prediction, conservation analysis, and structural characterization of mammalian mucin-type O-glycosylation sites | Glycobiology | Oxford Academic n.d. https://academic.oup.com/glycob/article/15/2/153/568173?login=false (accessed June 2, 2024).

[46] Chen Y-Z, Tang Y-R, Sheng Z-Y, Zhang Z. Prediction of mucin-type O-glycosylation sites in mammalian proteins using the composition of k-spaced amino acid pairs. BMC Bioinformatics 2008;9:101. https://doi.org/10.1186/1471-2105-9-101.

[47] Wandall HH, Nielsen MAI, King-Smith S, de Haan N, Bagdonaite I. Global functions of O-glycosylation: Promises and challenges in O-glycobiology. FEBS J 2021;288:7183–212. https://doi.org/10.1111/febs.16148.

[48] Edwards E, Livanos M, Krueger A, Dell A, Haslam SM, Mark Smales C, et al. Strategies to control therapeutic antibody glycosylation during bioprocessing: Synthesis and separation. Biotechnol Bioeng 2022;119:1343–58. https://doi.org/10.1002/bit.28066.

[49] Clausen H, Wandall HH, DeLisa MP, Stanley P, Schnaar RL. Glycosylation engineering. 2022.

[50] Prabhu A, Shanmugam D, Gadgil M. Engineering nucleotide sugar synthesis pathways for independent and simultaneous modulation of N-glycan galactosylation and fucosylation in CHO cells. Metab Eng 2022;74:61–71. https://doi.org/10.1016/j.ymben.2022.09.003.

[51] Kightlinger W, Warfel KF, DeLisa MP, Jewett MC. Synthetic glycobiology: Parts, systems, and applications. ACS Synth Biol 2020;9:1534–62. https://doi.org/10.1021/acssynbio.0c00210.

[52] Sealover NR, Davis AM, Brooks JK, George HJ, Kayser KJ, Lin N. Engineering Chinese Hamster Ovary (CHO) cells for producing recombinant proteins with simple glycoforms by zinc-finger nuclease (ZFN) – mediated gene knockout of mannosyl (alpha-1,3-)-glycoprotein beta-1,2-N-acetylglucosaminyltransferase (*Mgat1*). J Biotechnol 2013;167:24–32. https://doi.org/10.1016/j.jbiotec.2013.06.006.

[53] Zhong X, Cooley C, Seth N, Juo ZS, Presman E, Resendes N, et al. Engineering novel Lec1 glycosylation mutants in CHO–DUKX cells: Molecular insights and effector modulation of N-acetylglucosaminyltransferase I. Biotechnol Bioeng 2012;109:1723–34. https://doi.org/10.1002/bit.24448.

[54] Meuris L, Santens F, Elson G, Festjens N, Boone M, Dos Santos A, et al. GlycoDelete engineering of mammalian cells simplifies N-glycosylation of recombinant proteins. Nat Biotechnol 2014;32:485–89. https://doi.org/10.1038/nbt.2885.

[55] Klingler F, Naumann L, Schlossbauer P, Dreyer L, Burkhart M, Handrick R, et al. A novel system for glycosylation engineering by natural and artificial miRNAs. Metab Eng 2023;77:53–63. https://doi.org/10.1016/j.ymben.2023.03.004.

[56] Schweickert PG, Cheng Z. Application of genetic engineering in biotherapeutics development. J Pharm Innov 2020;15:232–54. https://doi.org/10.1007/s12247-019-09411-6.

[57] Zhang M, Koskie K, Ross JS, Kayser KJ, Caple MV. Enhancing glycoprotein sialylation by targeted gene silencing in mammalian cells. Biotechnol Bioeng 2010;105:1094–105. https://doi.org/10.1002/bit.22633.

[58] Wang Q, Stuczynski M, Gao Y, Betenbaugh MJ. Strategies for Engineering Protein N-Glycosylation Pathways in Mammalian Cells. In: Castilho A, editor. Glyco-Eng. Methods Protoc., New York, NY: Springer New York; 2015, pp. 287–305. https://doi.org/10.1007/978-1-4939-2760-9_20.

[59] Al-Rubeai M, editor. Cell Engineering: Glycosylation, vol. 3, Dordrecht: Springer Netherlands; 2002. https://doi.org/10.1007/0-306-47525-1.

[60] Nason R, Büll C, Konstantinidi A, Sun L, Ye Z, Halim A, et al. Display of the human mucinome with defined O-glycans by gene engineered cells. Nat Commun 2021;12:4070. https://doi.org/10.1038/s41467-021-24366-4.

[61] Chang MM, Gaidukov L, Jung G, Tseng WA, Scarcelli JJ, Cornell R, et al. Small-molecule control of antibody N-glycosylation in engineered mammalian cells. Nat Chem Biol 2019;15:730–36. https://doi.org/10.1038/s41589-019-0288-4.

[62] Wong NSC, Wati L, Nissom PM, Feng HT, Lee MM, Yap MGS. An investigation of intracellular glycosylation activities in CHO cells: Effects of nucleotide sugar precursor feeding. Biotechnol Bioeng 2010;107:321–36. https://doi.org/10.1002/bit.22812.

[63] Reinl T, Grammel N, Kandzia S, Grabenhorst E, Conradt HS. Golgi engineering of CHO cells by targeted integration of glycosyltransferases leads to the expression of novel Asn-linked oligosaccharide structures at secretory glycoproteins. BMC Proc 2013;7:P84, Springer.

[64] Yang Q, An Y, Zhu S, Zhang R, Loke CM, Cipollo JF, et al. Glycan remodeling of human erythropoietin (EPO) through combined mammalian cell engineering and chemoenzymatic transglycosylation. ACS Chem Biol 2017;12:1665–73.

[65] Prati EG, Matasci M, Suter TB, Dinter A, Sburlati AR, Bailey JE. Engineering of coordinated up-and down-regulation of two glycosyltransferases of the o-glycosylation pathway in Chinese hamster ovary (CHO) cells. Biotechnol Bioeng 2000;68:239–44.

[66] Dinter A, Zeng S, Berger B, Berger EG. Glycosylation engineering in Chinese hamster ovary cells using tricistronic vectors. Biotechnol Lett 2000;22:25–30.

[67] Heffner KM, Wang Q, Hizal DB, Can Ö, Betenbaugh MJ. Glycoengineering of mammalian expression systems on a cellular level. Adv Glycobiotechnology 2021;37–69.

[68] Brücher BL, Jamall IS. Cell-cell communication in the tumor microenvironment, carcinogenesis, and anticancer treatment. Cell Physiol Biochem 2014;34:213–43.

[69] Zhang X, Zhang H, Gu J, Zhang J, Shi H, Qian H, et al. Engineered extracellular vesicles for cancer therapy. Adv Mater 2021;33:2005709.

[70] Kalluri R, McAndrews KM. The role of extracellular vesicles in cancer. Cell 2023;186:1610–26.

[71] Engineering Extracellular Vesicles as Delivery Systems in Therapeutic Applications – Wang – 2023 – Advanced Science – Wiley Online Library n.d. https://onlinelibrary.wiley.com/doi/10.1002/advs. 202300552 (accessed June 6, 2024).

[72] Kojima R, Bojar D, Rizzi G, Hamri GC-E, El-Baba MD, Saxena P, et al. Designer exosomes produced by implanted cells intracerebrally deliver therapeutic cargo for Parkinson's disease treatment. Nat Commun 2018;9:1305.

[73] Ji P, Yang Z, Li H, Wei M, Yang G, Xing H, et al. Smart exosomes with lymph node homing and immune-amplifying capacities for enhanced immunotherapy of metastatic breast cancer. Mol Ther-Nucleic Acids 2021;26:987–96.

[74] Tian Y, Li S, Song J, Ji T, Zhu M, Anderson GJ, et al. A doxorubicin delivery platform using engineered natural membrane vesicle exosomes for targeted tumor therapy. Biomaterials 2014;35:2383–90. https://doi.org/10.1016/j.biomaterials.2013.11.083.

[75] Kojima R, Aubel D, Fussenegger M. Novel theranostic agents for next-generation personalized medicine: Small molecules, nanoparticles, and engineered mammalian cells. Curr Opin Chem Biol 2015;28:29–38.

[76] Scheller L, Fussenegger M. From synthetic biology to human therapy: Engineered mammalian cells. Curr Opin Biotechnol 2019;58:108–16.

[77] Dwyer RM, Ryan J, Havelin RJ, Morris JC, Miller BW, Liu Z, et al. Mesenchymal stem cell (MSC) mediated delivery of the Sodium Iodide Symporter (NIS) supports radionuclide imaging and treatment of breast cancer. Stem Cells Dayton Ohio 2011;29:1149–57. https://doi.org/10.1002/stem.665.

[78] Schug C, Urnauer S, Jaeckel C, Schmohl KA, Tutter M, Steiger K, et al. TGFB1-driven mesenchymal stem cell-mediated NIS gene transfer. 2019. https://doi.org/10.1530/ERC-18-0173.

[79]   Bashor CJ, Hilton IB, Bandukwala H, Smith DM, Veiseh O. Engineering the next generation of cell-based therapeutics. Nat Rev Drug Discov 2022;21:655–75. https://doi.org/10.1038/s41573-022-00476-6.

[80]   Davies A, Sage B, Kolluri K, Alrifai D, Graham R, Weil B, et al. TACTICAL: A phase I/II trial to assess the safety and efficacy of MSCTRAIL in the treatment of metastatic lung adenocarcinoma. J Clin Oncol 2019;37:TPS9116–TPS9116. https://doi.org/10.1200/JCO.2019.37.15_suppl.TPS9116.

[81]   Cannon MJ, Block MS, Morehead LC, Knutson KL. The evolving clinical landscape for dendritic cell vaccines and cancer immunotherapy. Immunotherapy 2019;11:75–79. https://doi.org/10.2217/imt-2018-0129.

[82]   Gardner A, de Mingo Pulido Á, Ruffell B. Dendritic cells and their role in immunotherapy. Front Immunol 2020;11:924. https://doi.org/10.3389/fimmu.2020.00924.

[83]   Pfeiffer IA, Hoyer S, Gerer KF, Voll RE, Knippertz I, Gückel E, et al. Triggering of NF-κB in cytokine-matured human DCs generates superior DCs for T-cell priming in cancer immunotherapy. Eur J Immunol 2014;44:3413–28. https://doi.org/10.1002/eji.201344417.

[84]   Koch EAT, Schaft N, Kummer M, Berking C, Schuler G, Hasumi K, et al. A one-armed phase I dose escalation trial design: Personalized vaccination with IKKβ-matured, RNA-loaded dendritic cells for metastatic uveal melanoma. Front Immunol 2022;13. https://doi.org/10.3389/fimmu.2022.785231.

[85]   Piper K, DePledge L, Karsy M, Cobbs C. Glioma stem cells as immunotherapeutic targets: Advancements and challenges. Front Oncol 2021;11. https://doi.org/10.3389/fonc.2021.615704.

[86]   Mohammed S, Sukumaran S, Bajgain P, Watanabe N, Heslop HE, Rooney CM, et al. Improving chimeric antigen receptor-modified T cell function by reversing the immunosuppressive tumor microenvironment of pancreatic cancer. Mol Ther 2017;25:249–58.

[87]   Savanur MA, Weinstein-Marom H, Gross G. Implementing logic gates for safer immunotherapy of cancer. Front Immunol 2021;12. https://doi.org/10.3389/fimmu.2021.780399.

[88]   Caliendo F, Dukhinova M, Siciliano V. Engineered cell-based therapeutics: Synthetic biology meets immunology. Front Bioeng Biotechnol 2019;7:43. https://doi.org/10.3389/fbioe.2019.00043.

[89]   Hong M, Clubb JD, Chen YY. Engineering CAR-T cells for next-generation cancer therapy. Cancer Cell 2020;38:473–88. https://doi.org/10.1016/j.ccell.2020.07.005.

[90]   Altamirano C, Berrios J, Vergara M, Becerra S. Advances in improving mammalian cells metabolism for recombinant protein production. Electron J Biotechnol 2013;16:10–10.

[91]   Luisa B. Cellular energy metabolism and its regulation. n.d.

[92]   Tanner LB, Goglia AG, Wei MH, Sehgal T, Parsons LR, Park JO, et al. Four key steps control glycolytic flux in mammalian cells. Cell Syst 2018;7:49–62, e8. https://doi.org/10.1016/j.cels.2018.06.003.

[93]   Arnold PK, Finley LWS. Regulation and function of the mammalian tricarboxylic acid cycle. J Biol Chem 2023;299:102838. https://doi.org/10.1016/j.jbc.2022.102838.

[94]   Edwards M, Mohiuddin S. Biochemistry, lipolysis. StatPearls n.d.

[95]   Torres N, Tobón-Cornejo S, Velazquez-Villegas LA, Noriega LG, Alemán-Escondrillas G, Tovar AR. Amino acid catabolism: An overlooked area of metabolism. Nutrients 2023;15:3378. https://doi.org/10.3390/nu15153378.

[96]   Hatting M, Tavares CDJ, Sharabi K, Rines AK, Puigserver P. Insulin regulation of gluconeogenesis. Ann N Y Acad Sci 2018;1411:21–35. https://doi.org/10.1111/nyas.13435.

[97]   Stangherlin A, Seinkmane E, O'Neill JS. Understanding circadian regulation of mammalian cell function, protein homeostasis, and metabolism. Curr Opin Syst Biol 2021;28:100391. https://doi.org/10.1016/j.coisb.2021.100391.

[98]   Otto AM. Warburg effect(s) – A biographical sketch of Otto Warburg and his impacts on tumor metabolism. Cancer Metab 2016;4:5. https://doi.org/10.1186/s40170-016-0145-9.

[99]   Saxton RA, Sabatini DM. mTOR signaling in growth, metabolism, and disease. Cell 2017;168:960–76. https://doi.org/10.1016/j.cell.2017.02.004.

[100] Maynard A, McCoach CE, Rotow JK, Harris L, Haderk F, Kerr DL, et al. Therapy-induced evolution of human lung cancer revealed by single-cell RNA sequencing. Cell 2020;182:1232–51, e22. https://doi.org/10.1016/j.cell.2020.07.017.

[101] Vander Heiden MG, DeBerardinis RJ. Understanding the intersections between metabolism and cancer biology. Cell 2017;168:657–69. https://doi.org/10.1016/j.cell.2016.12.039.

[102] Biancur DE, Kapner KS, Yamamoto K, Banh RS, Neggers JE, Sohn ASW, et al. Functional genomics identifies metabolic vulnerabilities in pancreatic cancer. Cell Metab 2021;33:199–210, e8. https://doi.org/10.1016/j.cmet.2020.10.018.

[103] Martínez-Reyes I, Chandel NS. Cancer metabolism: Looking forward. Nat Rev Cancer 2021;21:669–80. https://doi.org/10.1038/s41568-021-00378-6.

[104] Rathore R, Schutt CR, Van Tine BA. PHGDH as a mechanism for resistance in metabolically-driven cancers. Cancer Drug Resist 2020. https://doi.org/10.20517/cdr.2020.46.

[105] Mulukutla BC, Khan S, Lange A, Hu W-S. Glucose metabolism in mammalian cell culture: New insights for tweaking vintage pathways. Trends Biotechnol 2010;28:476–84. https://doi.org/10.1016/j.tibtech.2010.06.005.

[106] Templeton N, Young JD. Biochemical and metabolic engineering approaches to enhance production of therapeutic proteins in animal cell cultures. Biochem Eng J 2018;136:40–50. https://doi.org/10.1016/j.bej.2018.04.008.

[107] Funk CR, Wang S, Chen KZ, Waller A, Sharma A, Edgar CL, et al. PI3Kδ/γ inhibition promotes human CART cell epigenetic and metabolic reprogramming to enhance antitumor cytotoxicity. Blood 2022;139:523–37. https://doi.org/10.1182/blood.2021011597.

[108] Slattery K, Gardiner CM. NK cell metabolism and TGFβ – implications for immunotherapy. Front Immunol 2019;10. https://doi.org/10.3389/fimmu.2019.02915.

[109] Mitchell DA, Batich KA, Gunn MD, Huang M-N, Sanchez-Perez L, Nair SK, et al. Tetanus toxoid and CCL3 improve DC vaccines in mice and glioblastoma patients. Nature 2015;519:366–69. https://doi.org/10.1038/nature14320.

[110] Xia C, Dong X, Li H, Cao M, Sun D, He S, et al. Cancer statistics in China and United States, 2022: Profiles, trends, and determinants. Chin Med J (Engl) 2022;135:584–90. https://doi.org/10.1097/CM9.0000000000002108.

[111] Pająk B, Zieliński R, Priebe W. The impact of glycolysis and its inhibitors on the immune response to inflammation and autoimmunity. Molecules 2024;29:1298. https://doi.org/10.3390/molecules29061298.

[112] Phan TG, Croucher PI. The dormant cancer cell life cycle. Nat Rev Cancer 2020;20:398–411. https://doi.org/10.1038/s41568-020-0263-0.

[113] Orman MA, Androulakis IP, Berthiaume F, Ierapetritou MG. Metabolic network analysis of perfused livers under fed and fasted states: Incorporating thermodynamic and futile-cycle-associated regulatory constraints. J Theor Biol 2012;293:101–10. https://doi.org/10.1016/j.jtbi.2011.10.019.

[114] Artificial signaling in mammalian cells enabled by prokaryotic two-component system | Nature Chemical Biology n.d. https://www.nature.com/articles/s41589-019-0429-9 (accessed June 2, 2024).

[115] Beránek V, Willis JCW, Chin JW. An evolved *Methanomethylophilus alvus* pyrrolysyl-tRNA synthetase/tRNA pair is highly active and orthogonal in mammalian cells. Biochemistry 2019;58:387–90. https://doi.org/10.1021/acs.biochem.8b00808.

[116] MacDonald IC, Deans TL. Tools and applications in synthetic biology. Adv Drug Deliv Rev 2016;105:20–34. https://doi.org/10.1016/j.addr.2016.08.008.

Doğuş Akboğa, Aslı Doğruer, Damla Albayrak, Gozeel Binte Shahid,
and Urartu Özgür Şafak Şeker

Chapter 4
Engineering microbial cells for cancer

Abstract: The use of engineered microbes in personalized cancer treatment is promising, offering targeted and patient-specific therapeutic strategies. Advances in synthetic biology enable precise genetic modifications, creating bacteria capable of producing therapeutic agents within tumors. This chapter discusses the field of engineering microbial cells for cancer therapy, initially focusing on the dynamic interactions between the microbiome and the host and the pivotal role of the gut microbiome in cancer development. Then, discussing that certain bacteria influence cancer progression and therapeutic responses through immunomodulation and metabolic interactions, how engineered bacteria can present novel opportunities for intervention will be explored. The principles of bacterial cancer therapies will be examined, including selecting suitable bacterial strains and engineering methods to ensure their safety and efficacy. Techniques such as attenuating virulence factors, enhancing tumor-targeting capabilities, and designing sophisticated genetic circuits for controlled therapeutic delivery are detailed.

4.1 Introduction

Microorganisms display remarkable ecological adaptations, which allow them to thrive in every environment on the Earth, from the deepest ocean trenches to the highest mountain peaks. Their vast diversity has captivated human interest, sparking a long journey of studies to highlight their extensive presence and interactions within localized "microworlds." Microbiologists frequently use two terms to describe these microbial communities: "Microbiota" refers to the distinct microbial taxa associated with a host organism or dominant habitat, whereas "microbiome" encompasses the entire collection of these microorganisms and their related genes [1].

The complex and intricate human microbiome includes viruses, archaea, bacteria, and eukaryotic organisms that inhabit our bodies' surfaces and internal regions. The majority of them are commensal or mutualistic microorganisms [2]. This diverse community is crucial in maintaining health, influencing everything from digestion to immune function. Understanding the human microbiome extends beyond academic interest; it represents a frontier in medical science, offering profound insights and opportunities into disease prevention, diagnosis, and treatment.

Doğuş Akboğa, Aslı Doğruer, Damla Albayrak, Gozeel Binte Shahid, Urartu Özgür Şafak Şeker,
UNAM–Institute of Materials Science and Nanotechnology, Bilkent University, Ankara 06800, Turkey

https://doi.org/10.1515/9783111329499-004

4.1.1 Microbiome–host interactions

The microbiome performs critical roles in the human body, such as barrier modulation, homeostasis maintenance, pathogenic infection prevention, and metabolic and vitamin synthesis management [3]. It has a vast and diversified microbial ecosystem that evolved with our species and plays a significant part in human health. This symbiotic relationship is manifested through proteomic and metabolic interactions inside the host, leading to the hypothesis that humans can be viewed as a sophisticated biological "superorganism" that shares some metabolic regulatory functions with microbial symbionts. The oral, gut, and skin microbiomes have abundant and diverse communities of microorganisms, whereas the lungs, bladder, prostate, liver, pancreas, and vagina have less diversified microbial populations [4]. This symbiotic superorganism offers both entities a nutrient-rich milieu [5].

Variability in microbial diversity and abundance throughout different organs are associated with disease occurrence, with some organs being more vulnerable than others. For example, specific disorders like cancer are more likely to originate in areas with high microbial densities [6]. Also, certain bacteria significantly impact human physiology by maintaining metabolic functions, offering protection against infections, refining the immunological response, and exerting a direct or indirect influence on most physiological processes. The gastrointestinal (GI) tract (GIT) is the primary contact between the environment, the host, and the host's body antigens, which, according to some estimates, hosts around 10^{14} bacteria, outnumbering human cells roughly tenfold and possessing a microbial genetic content that exceeds the complete human genome by more than 100 times [7–9]. This extensive microbial diversity extends the functional capabilities of the host genome, integrating a vast array of microbial genes that enhance overall complexity and functionality. These genomes add to an expanded repertoire of enzymes the host does not typically encode, which is essential in enabling host metabolism and contributing to host's physiological regulation [10, 11].

Despite the richness of microbial diversity, technological limitations, especially in studying noncultivable microbes, and the need for more accurate information on the functions and compositions of the microbiota have historically obscured the understanding of host–microbiota properties and interactions within the human microbiome. However, advanced sequencing techniques, such as next-generation sequencing, have significantly improved our ability to analyze microbial communities. The decreasing cost and increased speed of DNA sequencing, combined with advances in computational tools for processing large datasets, have enabled extensive research into bacterial populations living on or within the human body [12–14]. Extensive sequence-based initiatives such as the National Institutes of Health (NIH)-funded Human Microbiome Project (HMP) and the European Commission, and CORDIS-funded Metagenomics of the Human Intestinal Tract (MetaHIT) Consortium have accelerated the gut microbiome research in recent years [15–17]. These initiatives have conducted comprehensive surveys of bacterial populations using small-subunit (16S) ribosomal RNA gene sequences, which are present in all microorganisms and provide essential sequence conservation and variation for phylogenetic analysis. These surveys analyzed various bodily locations, including the skin, mouth, esophagus, stomach, colon, and vagina [18–23].

Moreover, next-generation "omics" tools supported by systems biology efforts have transformed our understanding of the gut microbiome by enabling deep genetic and functional investigations at the transcriptomic, proteomic, and metabolic levels [24–26]. Therefore, modern understanding of the gut microbiome is based on molecular approaches rather than traditional culture-based methods, aided by high-throughput genomic screening technology. These improved perspectives and aspects have shed new light on the crucial function of the gut microbiome in human health by revealing microbiome heterogeneity among animals, individuals, and groups. Studies have emphasized the gut microbiome's role in developing various systemic disorders, including obesity, cardiovascular disease, and intestinal conditions like inflammatory bowel disease [27–30]. Furthermore, the gut flora was found to be influencing drug metabolism, dietary calorie absorption, immune system modulation, and postsurgical healing [28, 31–36]. As a result, understanding the microbiome's activity is vital for developing future customized healthcare strategies and identifying novel medication targets.

4.1.2 Gut microbiome and cancer

In the late nineteenth century, Wilhelm Busch [37] and Friedrich Fehleisen [38] observed spontaneous tumor regressions in patients with *Streptococcus pyogenes* infections. Building on this and other notable reports of tumor regressions in cases of infectious diseases, William Coley developed a vaccine using a mixture of live or heat-killed *Streptococcus* and *Serratia* species for terminal cancer patients, performing the first intentional use of immunotherapy in 1891 [39]. Concurrently, Thomas Glover and Virginia Livingston-Wheeler proposed, controversially, that bacteria could be cultured from tumors and that bacterial vaccines were effective against cancer. However, their claims of the bacterial theory of cancer were eventually dismissed due to irreproducible results, lack of mechanistic evidence, and the hazardous nature of their treatments [40].

The characterization and establishment of the hallmarks of human cancer, such as sustaining proliferation, evading growth suppression, activating invasion and metastasis, and inducing angiogenesis, combined with the recognition that the human body contains as many microbial cells as human cells with microbial genes vastly outnumbering human genes, have shifted perspectives on the role of microbes in cancer diagnosis, pathogenesis, and treatment [41–45]. Microbial niches can influence cancer promotion through dysbiosis, direct interactions, or secreted metabolites, and they can significantly impact host immune system development and antitumor immunosurveillance [46–50]. Research also linked the gut microbiota's role in modulating responses to cancer immunotherapy and the impact of microbial communities within the tumor microenvironment (TME) on therapeutic efficacy [51]. While much research has focused on gut microbiota, recent studies have highlighted the presence, metabolic activity, and importance of intratumoral microbiota [52–55]. The complexity of the interactions likely reflects the evolutionary dynamics between the host's immune system, its commensal microbiota, and tumorigenic processes. The impact of the human microbiome on cancer progression and therapy and its interactions with the immune system and TME are not the main focuses of this chapter, as there are extensive reviews that undertake that endeavor [42–44, 51, 56]. However, some fundamental knowledge on this topic will be provided before progressing into how engineered bacterial therapies take advantage of these interactions.

4.1.2.1 Role of microbes in cancer development

While only a few microbes directly cause cancer, many more play a supportive role, and some even enhance antitumor immunity [42, 57]. It is essential to understand the distribution of microbes in the body, as approximately 4×10^{13} microbial cells, covering around 3,000 species, exist within us, with most being colonic bacteria responsible for the known immunomodulatory effects [7, 58]. Of the estimated $\sim 10^{12}$ distinct microbial species on the Earth, only 11 are recognized as human carcinogens by the International Association of Cancer Registries (IACR): the *Epstein–Barr* virus, *hepatitis B* virus, *hepatitis C virus, Kaposi's sarcoma-associated herpesvirus, human immunodeficiency virus-1, human papillomaviruses, human T-cell lymphotropic virus type 1, Opisthorchis viverrini, Clonorchis sinensis, Schistosoma haematobium*, and *Helicobacter pylori* [59].

H. pylori can have direct genotoxic effects and alter critical intracellular signaling pathways. These bacteria, found in over half of the global population, are known to cause peptic ulcers, gastric cancers, and mucosa-associated lymphoid tissue lymphomas by disrupting Wnt/β-catenin pathways and triggering chronic inflammation through proinflammatory signaling [60–63]. This bacterium attaches to gastric epithelial cells using the outer membrane adhesin HopQ to bind to carcinoembryonic antigen-related cell adhesion molecules, then injects CagA, which directly interacts with E-cadherin and disrupts the E-cadherin-β-catenin association, into the cells via the type 4 secretion system [64–68]. The disruption leads to nuclear β-catenin accumulation. The delivered CagA acts as a scaffold protein, interacting with SHP2, PAR1/MARK, and PI3K, dependent and independent of tyrosine phosphorylation, which promotes the neoplastic transformation of gastric epithelial cells [69].

Beyond *H. pylori*, recent studies have identified many microbial species that can influence or contribute to cancer [54, 70–75]. While not directly causing cancer, these "complicit" microbes promote carcinogenesis through their immunomodulatory functions and bioactive metabolites. They may aid tumor progression through feedback loops between tumors and surrounding tissues, inflammation, or compromised immunosurveillance. For example, colibactin, a toxin from pathogenic *E. coli*, causes DNA damage, leading to mutations in colorectal, head, neck, and urinary tract cancers [70, 76]. Similarly, *enterotoxigenic Bacteroides fragilis* promotes colitis and colorectal cancer (CRC) by producing a toxin that increases inflammation and intestinal permeability [77]. *Fusobacterium nucleatum* is linked to CRC and liver metastasis, promoting tumor progression through specific signaling pathways and inducing proinflammatory cascades [74, 78–80]. Other microbes, such as *Salmonella*, activate pathways that sustain cellular transformation and manipulate immune responses to support tumor growth. For instance, the injection of AvrA from *Salmonella enterica* enhances β-catenin pathways, activating STAT3 signaling, inflammation, and epithelial–mesenchymal transition (EMT)-inducing transcription factors [81, 82].

4.1.2.2 Interactions between microbiome and immune system and TME

Shifts in the microbiome within the lower GIT are strongly linked to GI cancers, particularly CRC [83]. Research from preclinical and human studies suggests that gut dysbiosis is a causal factor in CRC. Unlike the adjacent healthy mucosa, tumor microbiota can induce polyp formation, trigger procarcinogenic signals, and alter the local immune environment [84]. A healthy gut microbiome typically includes species such as *Lactobacillus*, *Bacteroides*, and *Bifidobacterium*. However, in CRC, there is often an overrepresentation of *Fusobacterium, Porphyromonas, Parvimonas, Peptostreptococcus*, and *Gemella* spp., indicating microbial imbalance or dysbiosis as colorectal tumors progress from adenomas to carcinomas, distinct microhabitats with unique microbial communities emerge, including an increased presence of oral-associated microbes like *Fusobacteria* in colorectal lesions [84, 85].

While the role of live microbiota in various tumor types is generally limited outside the aerodigestive tract, there are intratumor microbes that can suppress local antitumor immunity. For instance, intratumor microbes in the urinary tract secrete genotoxins, influencing chemoresistance, tumor growth, and metastasis [54, 86–90]. These microbes can create a tolerogenic environment by engaging pattern recognition receptors (PRRs), reducing tumor-infiltrating lymphocytes, including CD8+ T cells, and sometimes increasing regulatory T cells (Tregs). Conversely, intratumoral bacteria or their antigens can occasionally stimulate the immune system, as demonstrated by Coley's toxins and recent bacterial cancer therapies [39, 91].

Microbial signals can also influence the tumor-associated adaptive immune response. T-cell exhaustion within the TME can be affected by microbial-secreted metabolites, outer membrane vesicles (OMVs), or intratumoral bacteria, creating various immunomodulatory effects that support tumor growth and create immunosuppressive environments [92–94]. Bacterial microbe-associated molecular patterns can stimulate antitumor immunity through PRR signaling. Gut microbiome-modulated bioactive metabolites can impact tumor immunity, affecting innate and adaptive immune responses. For example, *B. fragilis* OMVs carrying polysaccharide A exhibit anti-inflammatory properties that protect against CRC proliferation and suppress EMT [95].

4.1.2.3 Impact of microbiome on cancer therapy

The microbiome significantly influences cancer therapy by either enhancing or hindering treatment efficacy. Microbiota-derived compounds, like propionate and tryptophan metabolites, help protect against radiation-induced damage to the hematopoietic system by supporting bone marrow-derived myeloid cells and neutrophil function [96, 97]. This protection may involve the delivery of endogenous ligands for RIG-I, which induce type I interferon (IFN) signaling, aiding in the repair of the intestinal barrier [98].

Certain commensal bacteria are strongly associated with protective antitumor T-cell responses during cancer treatments. For instance, cyclophosphamide promotes the translocation of *Enterococcus hirae*, which in turn stimulates helper T-cell 17 (TH17) responses and IFN-producing CD8+ T-cell effectors, effectively curbing tumor growth in sarcoma and lung adenocarcinoma models [99, 100]. Similarly, in melanoma patients, CTLA-4 blockade increases the presence of *Bacteroides thetaiotaomicron* and *B. fragilis*, enhancing therapeutic efficacy through toll-like receptor 4 (TLR4) and interleukin (IL)-12-dependent TH1 responses [101].

The gut microbiota can act as an adjuvant in cancer therapy by stimulating systemic immune responses. Dendritic cells (DCs) from gut-associated lymphoid tissue, spleen, or tumor-draining lymph nodes can sense various commensals, such as *Bifidobacterium* spp., *B. fragilis*, and *Akkermansia muciniphila* [102–104]. These interactions catalyze immune responses through IFN-I- and IL-12-mediated pathways, enhancing antitumor immunity and providing systemic benefits beyond the gut environment.

4.1.2.4 Bacterial cancer therapy

As mentioned before in this chapter, the use of bacteria in cancer therapy dates back to the nineteenth century when physicians like Wilhelm Busch and William Coley discovered that live bacteria could potentially shrink tumors. These early discoveries laid the foundation for modern bacterial therapies. Fecal microbiota transplantation (FMT), which involves introducing stool from a healthy donor into a patient's GI system, has gained attention as a therapeutic approach. Approved initially to treat antibiotic-resistant *Clostridioides difficile* infections, FMT is now being explored for cancer treatment due to its potential impact on gut microorganisms like *F. nucleatum*, *B. fragilis*, and *Enterococcus faecalis* [105, 106]. Studies have shown that FMT can enhance antitumor responses, particularly when combined with PD-1 blockade [107–109].

In addition to FMT, using prebiotics, probiotics, and dietary interventions to modify the microbiome has shown promise in cancer therapy. Prebiotics help promote the growth of beneficial microbes, while probiotics introduce beneficial bacteria into the gut. These interventions can boost antitumor immunity and enhance the effectiveness of cancer treatments. For instance, dietary changes and specific microbial supplements have been found to support immune function and improve responses to therapies like checkpoint inhibitors [110].

Synthetic biology has revolutionized bacterial cancer therapy by enabling precise genetic modifications of bacterial cells to create live bacterial therapeutics (LBTs) [111–113]. This field involves designing and constructing new biological parts, devices, and systems or redesigning existing biological systems for practical purposes. With synthetic biology, scientists can engineer bacteria to perform specific tasks within the TME, such as producing and releasing therapeutic agents, targeting tumor cells, and stimulating the immune system.

Advancements in synthetic biology have significantly transformed bacterial cancer therapy. Genetically modified bacteria can now be used as "intratumoral bioreactors" that continuously produce and release therapeutic compounds within the tumor, ensuring a high local concentration of the therapeutic agent [42]. This approach improves treatment effectiveness and reduces side effects. Moreover, bacteria can be engineered to lyse and release their payloads at specific population densities, further refining treatment precision. These innovations highlight the transformative potential of synthetic biology in developing safe and effective bacterial therapies for cancer. The following section explores strategies to program bacterial cells for cancer treatment, focusing on specific tools and applications.

4.2 Principles of engineering bacterial cancer therapies

Significant progress has been made in developing bacterial therapies for various cancers in recent years. These therapies offer promising alternatives for patients with previously untreatable malignancies. Bacterial therapies can colonize primary and metastatic tumors, deliver therapeutic agents directly to tumor cells, induce direct cancer cell death, and stimulate immune responses against tumors. They provide advantages over traditional treatments like chemotherapy by targeting malignant tissues specifically, thus reducing systemic toxicities. Moreover, they can address tumors unsuitable for surgery or radiation and bypass the patient specificity challenges seen with chimeric antigen receptor T-cell therapy and viral cancer vaccines. This focused delivery, with low systemic side effects and potential effectiveness against hard-to-treat cancers, positions bacterial therapeutics as a promising frontier in cancer treatment. The selection of appropriate bacterial chassis is critical for building effective bacterial cancer therapies, which will be discussed next [114, 115].

4.2.1 Bacterial host selection

Insights into bacterial activity within tumors have been gained by analyzing their interactions with epithelial and immune cells in the intestines, as both tumors and intestines feature microbially favorable traits [116]. Understanding microbial infection pathways is critical for creating enhanced bacterial cancer treatments. Similar nutritional gradients affect microbial development in both malignancies and the gut. For example, fermentable fibers are abundant in the ascending colon's lumen, but oxygen levels are low. Tumors exhibit similar gradients of nutrients and oxygen. Viable tumor tissue and active immune cells are usually found near the tumor vasculature, with higher nutrition and oxygen levels. Tumor regions far from the vasculature undergo hypoxia and are predominantly made up of dead cells and debris, resulting in immune-privileged settings permissive to bacterial development due to the lack of functional immune cells. Recent research has shown that bacterial species from the intestinal microbiome can naturally colonize tumor sites, influencing the efficacy of immune-based therapy [117].

Salmonella, particularly *S. enterica* serovar *Typhimurium*, has emerged as a viable therapeutic vector for selectively delivering active compounds to tumors while limiting damage [115, 118]. It can infect both phagocytic and non-phagocytic cells, allowing compounds to be delivered into the intracellular and extracellular areas of malignancies. *Salmonella* therapy can target a diverse spectrum of cells in the TME, including epithelial cancer cells, tumor-associated immune cells, and stromal cells. *Salmonella* is a facultative anaerobe and pathogen that can infect the large intestine. *Salmonella* competes for resources with the microbiome within the intestinal lumen but avoids competition by crossing the mucosal barrier using flagella-dependent motility. Upon entering gut epithelial cells, *Salmonella* attaches to cell membranes using the *Salmonella* pathogenicity island-1 type 3 secretion system (T3SS), injecting effector proteins into cells, causing membrane rearrangement and endocytosis into intracellular vacuoles [119, 120].

Commensal *E. coli* strains colonize the GIT shortly after birth. Upon initial colonization, these bacteria consume oxygen and thrive aerobically. The absence of antimicrobial immune cells in the intestinal lumen promotes their development [121]. However, if *E. coli* crosses the mucosal barrier, they are ingested by phagocytic cells and presented by antigen-presenting cells (APCs) [122]. The intestinal milieu, metabolic adaptability, and absence of virulence factors promote *E. coli* colonization in the intestine. *E. coli* has demonstrated efficacy as a selective delivery vehicle for malignancies. Several genetic alterations have been made to improve its safety and delivery efficiency. Notably, variants of the gut commensal strain *E. coli* MG1655 and the probiotic strain

E. coli Nissle 1917 (EcN) have been extensively modified for cancer therapy, with the latter currently in clinical trials. EcN has been engineered to lyse and release medicinal compounds into the extracellular space or deliver stimulators of IFN gene (STING) agonists when taken up by APCs [123].

Probiotic microorganisms have shown potential in cancer treatment. Gut microbes can produce compounds that either promote cancer or alter the efficacy of immunological therapies. Several probiotic species, including *Bifidobacterium infantis* [124], *Bifidobacterium longum* [125], *Lactobacillus acidophilus* [126], *Lactobacillus rhamnosus* [125], and *Limosilactobacillus fermentum* [127], have been examined in this context. *Bifidobacterium*, for instance, has been used to transmit viral genes that activate prodrugs or show tumor antigens that sensitize the immune system. When delivered intravenously as spores, *Bifidobacterium* can selectively target and colonize hypoxic regions within tumors [124]. Furthermore, several probiotic species exhibit therapeutic properties in their natural form and are routinely used as food supplements or components in fecal transplants. Investigating the complex link between the microbiome and tumor growth has tremendous potential for improving cancer treatments. Bacteria have a unique advantage over other delivery systems as they can continuously and precisely administer a wide range of therapeutic agents, offering higher doses of therapy to tumors compared to healthy tissues due to their exponential proliferation within malignant tissues [115].

4.2.2 Engineering strategies

Microbial systems, particularly bacteria-based systems, have demonstrated immense potential in cancer therapy due to cutting-edge innovations in synthetic biology. The synthetic biology tools facilitate complex genetic modifications of these microbial hosts to precisely target tumors with minimal toxicity to the host body and deliver a variety of therapeutic agents that can act on the tumors directly or recruit immune cells to the TME to elicit antitumor immune responses [112, 128]. This section highlights the genetic engineering methods applied to bacteria for cancer therapy from several critical perspectives. First, the attenuation of bacterial strains to enhance the safety profile of the therapies and further genetic modifications to improve the tumor-targeting efficiency of these strains will be examined. Second, various sense and response elements in genetic circuits designed for cancer therapies will be explored. Finally, different approaches will be addressed to integrate these elements into complex genetic circuits in bacterial hosts.

4.2.2.1 Attenuations and genetic modifications of bacterial strains

The primary challenge in bacterial cancer therapeutics is the toxicity associated with the virulence factors intrinsic in certain bacterial strains. This issue can be addressed by attenuating these wild-type strains by introducing mutations or deletions to the virulence genes. This approach has successfully transformed fatally toxic bacterial strains into safe clinical research strains [129]. The *S. typhimurium* strain VNP20009 is one of the remarkable examples of attenuated strains [130]. It was modified by the deletion of the msbB and purI genes. Deleting msbB modifies the lipid A component of lipopolysaccharide, significantly reducing the production of tumor necrosis factor (TNF)-α. The other deletion in purI renders bacteria reliant on an exogenous source of purine, which is abundant in the TME (Figure 4.1A). Phase I trials of VNP20009 in metastatic melanoma patients showed that while the patients safely tolerated the treatment, it failed to colonize tumors efficiently [131].

The therapeutic effect of *S. typhimurium* was improved in another attenuated strain, which lacks the znuABC operon coding for a high-affinity zinc transporter. Deleting this operon confers selective proliferation in tumor tissues, where zinc concentration is higher than in normal tissues [132]. Further genetic modifications were explored in the attenuated *S. typhimurium* strain ΔppGpp, which is defective in guanosine 5′-diphosphate-3′-diphosphate production. This strain has demonstrated remarkable antitumor activity by enhancing the synthesis of proinflammatory cytokines IL-1β and TNF-α [133].

Genetic modifications of bacterial strains have the potential to improve not only their safety profile but also their antitumor activity. One effective strategy to enhance the antitumor activity of bacterial strains is introducing genetic mutations or deletions to enhance bacteria's intrinsic metabolic properties [134]. For instance, L-arginine is a crucial amino acid for T-cell survival and activity. *Escherichia coli* can synthesize L-arginine from ammonia, commonly accumulated in the TME due to the highly active amino acid metabolism [129]. Canale et al. [134] engineered the EcN strain to convert ammonia into L-arginine more efficiently. They deleted the arginine repressor gene ArgR and introduced the ArgAfbr gene that encodes for an arginine synthase enzyme resistant to the negative feedback from high levels of L-arginine (Figure 4.1B). The intratumoral injection of this engineered strain led to an increase in the tumor-infiltrating T cells and antitumor immune activity [134]. These modifications highlight the potential of genetically engineered bacterial strains in cancer therapy by providing promising methods to enhance the safety and effectiveness of these strains.

4.2.2.2 Sense and response elements

Synthetic biology tools enable the engineering of bacteria for cancer therapy by constructing genetic circuits that are capable of sensing specific signals and responding

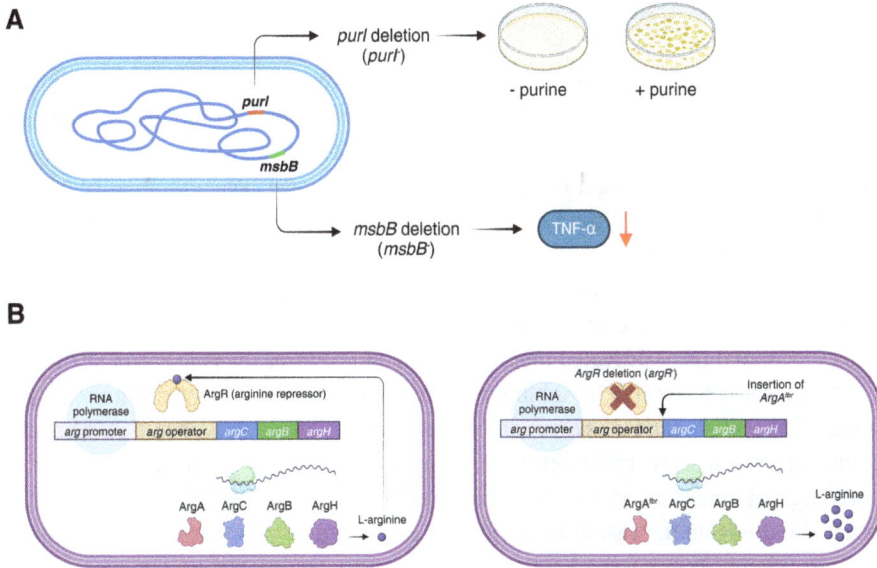

Figure 4.1: Schematic representation of attenuated bacteria genetically engineered for cancer therapy. (A) Genetically modified VNP20009 strain of *S. typhimurium* with improved safety profile, showing the deletions in the purI and msbB loci in the genome. **(B)** Metabolically engineered strain of EcN producing high levels of L-arginine in tumors.

with the delivery of therapeutic payloads such as cytotoxic proteins, tumor-targeting nanobodies, cytokines, and immunomodulators in an efficient and controllable manner [128]. The genetic circuits can be designed to sense and respond to external signals, internal signals from the TME, or self-signals [135].

The external signals can be classified into chemical inducers and physical signals. Chemical inducers facilitate the dose-dependent and timely synthesis and release of therapeutic payloads [128]. For example, pBAD vector systems in *S. typhimurium* induced by the chemical L-arabinose have been programmed to encode for proteins such as the FlaB subunit of the flagellin or the pore-forming toxin cytolysin A (ClyA) to achieve motility toward the tumor site and an antitumor immune response (Figure 4.2A) [136, 137]. Jiang et al. [138] developed an alternative inducible system containing the bidirectional pTet promoter (pTetA and pTetR) in *S. typhimurium*. Upon induction with tetracycline, the system expresses ClyA as a therapeutic agent and the bioluminescent reporter protein, Renilla luciferase 8 (Rluc8), to verify tumor colonization [138]. Although chemical inducers are widely used to control the timing of the expression in a dose-dependent manner, the leaky gene expressions in these systems and the risk of high-dose toxicity are the limitations of the chemical-inducible systems [139]. Leaky expression in the absence of chemical inducers can be reduced using leak dampener tools such as suppressor transfer RNAs, toehold switches as translational regulators and small transcription activating RNAs as transcriptional regulators, and viral proteases found in plants [140–142].

Physical cues such as optical and thermal signals are attractive alternative options for the external triggering of bacteria. Unlike chemical inducers, physical signals are nontoxic and administered noninvasively [139]. Optical induction has been achieved by Ganai et al. [143] by utilizing a radiation-responsive promoter, RecA, to produce a TNF-related apoptosis-inducing ligand in the VNP20009 strain of *S. typhimurium* (Figure 4.2B) [143]. As for thermal induction, orthogonal heat switching systems have been introduced to EcN to synthesize antitumor payloads such as TNF-α or melanin [144, 145]. Another thermal regulation system was developed by Abedi et al. [146] in EcN in which the focused ultrasound (FUS)-activated thermal stimulation releases the expression of CTLA-4 and PD-L1 immune checkpoint inhibitors from the repression of temperature-sensitive protein TcI42. While the FUS system and systemically delivered CTLA-4 and PD-L1 nanobodies achieved a comparable reduction in tumor size, the FUS system demonstrated a more targeted tumor localization. This enhanced selectivity of the FUS-actuated bacterial system can mitigate the risk of unintentional autoimmunity associated with the direct administration of immune checkpoint inhibitors [146].

In addition to external triggers, bacteria can be engineered to respond to tumor-specific internal signals such as low pH, hypoxic conditions, and glucose concentration in the TME. When tumors reach a specific volume, accumulation of metabolic waste products and insufficient oxygen supply create an acidic and hypoxic TME [147]. Furthermore, tumor tissues are characterized by lower glucose concentrations due to the Warburg effect. According to the Warburg effect, tumor cells and tumor-infiltrating immune cells rely on glycolysis for energy production, even in the presence of oxygen in the TME [129]. These properties of TME offer promising strategies to deliver therapeutic agents with engineered bacteria in response to distinct tumor signatures.

TME has a weakly acidic pH in the range of 5.8–6.6. Promoters responding to acidic conditions have been used to restrict the payload production and delivery to the TME [129]. For instance, Qin et al. [148] constructed a system in *E. coli* containing an acid-sensitive promoter, adiA, that responds to acidic TME and controls the expression of ClyA (Figure 4.2C) [148]. Another characteristic condition of the TME is the hypoxic niche, characterized by low oxygen concentrations. Facultative or obligate anaerobic bacteria such as the species from *Clostridium, Escherichia, Lactobacillus, Listeria,* and *Salmonella* genera can be good candidate drug carriers in the hypoxic intratumor space [129]. Harnessing the hypoxic condition of tumors, Chien et al. [149] introduced a multiplexed biosensor system to EcN. They used a hypoxia-sensing promoter, pPeptT, to create a hypoxia biosensor and paired it with a lactate biosensor that detects the end product of glycolysis. These circuits were coupled with an AND gate to ensure that low oxygen and high lactate levels are required to activate the system. Upon receiving both tumor-specific signals, the circuits triggered the expression of asd and glms genes that are essential for bacterial growth. This system showed an enhanced localization of EcN to tumor sites [149]. Other than sensing the metabolic end product, the dependence of tumor cells on glycolysis can be utilized differently. Panteli and Forbes [150] proposed a tumor-specific drug delivery approach based on detecting lower glucose levels at the tumor site. They programmed *E. coli* to encode a chemotaxis-osmoporin fusion protein, Trz1, to sense glucose and ribose in the tumor mass [150].

The triggering methods are limited to external and internal stimuli and the self-signaling molecules integral to bacteria's quorum-sensing (QS) system [135]. Bacteria can selectively colonize tumors compared to normal tissues due to hypoxic conditions and immune evasion. QS-based circuits can facilitate communication among growing bacteria through molecules such as acyl-homoserine lactone and control the therapeutic payload synthesis upon reaching a threshold cell density in the TME [112]. Self-signaling was used in the research of Chowdhury et al. [151], in which a QS-based vector system was constructed in *E. coli* to deliver nanobodies against CD47, an antiphagocytic receptor overexpressed in several types of human cancer. In the system, reaching a specific cell density induced the expression of phage-derived lysis protein, ϕX174E, and resulted in bacterial lysis through which CD47 nanobodies were released to intratumoral space. Localized blockade of CD47 on tumor cell surfaces enhanced tumor infiltration of T cells and achieved durable antitumor immunity in mouse models [151]. A similar QS system with ϕX174E-induced bacterial lysis was developed in EcN by Gurbatri et al. [152] to release CTLA-4 and PD-L1-blocking nanobodies to tumors. This system showed localized expression of therapeutic payloads and improved antitumor activity (Figure 4.2D) [152]. The QS-based strategies control the bacterial density and provide a sustained and targeted antitumor drug delivery at the site of tumors [151, 152]. However, despite the advantages of QS approaches, challenges such as the potential inability to reach the threshold density at specific tumor sizes should be considered while designing the drug delivery platforms. Additionally, using lysis genes derived from bacteriophages could pose an evolutionary pressure on bacteria to undergo mutations and alter the system's effectiveness.

Upon sensing the specific signals, bacteria can be engineered to respond with a variety of intracellular and extracellular drug delivery mechanisms by expressing genes responsible for tumor localization, different protein secretion systems, passive transport of the payloads from bacteria to the cytoplasm of the tumor cells, and bacterial lysis.

The synthetic genetic circuits for bacteria-based cancer therapies can be built by integrating the sensing above and response elements in different combinations. Boolean logic gates (AND, OR, XOR, NOT, NAND, etc.) facilitate the multiplexing approaches to build complex genetic circuit systems. Since specific antigens expressed on the surfaces of solid tumors and other cellular components in TME are also present in normal cells, targeting tumors with bacterial systems could pose safety risks. Achieving enhanced tumor-targeting and choosing optimal combinations of signals to induce therapeutic delivery are the main challenges in bacterial cancer therapies. Constructing Boolean logic gates, especially "AND" gates that require the presence of all input signals to activate the output signal, can improve tumor targeting and mitigate the risk of

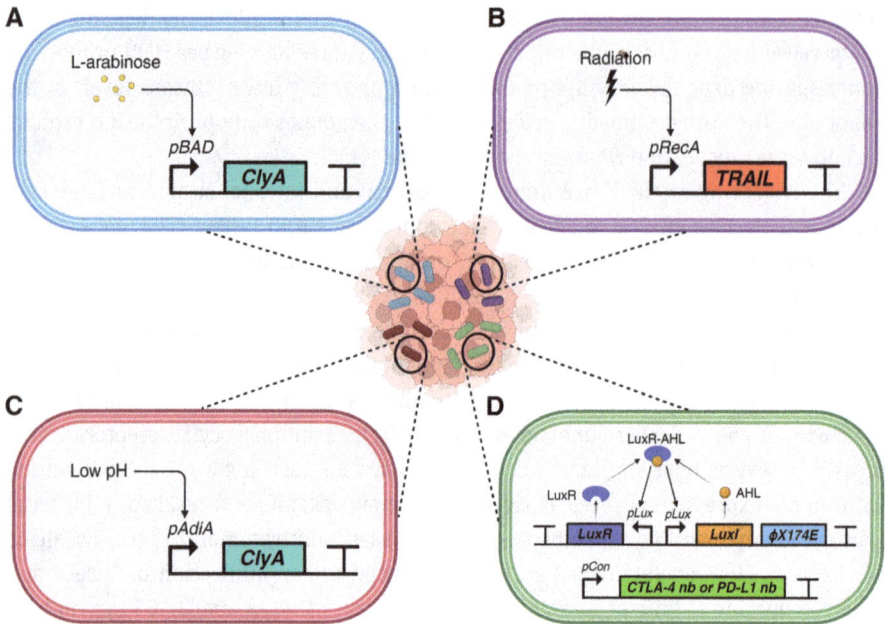

Figure 4.2: Genetic elements sensing and responding to external signals (A and B), TME-associated internal signals (C), and QS-based self-signals (D). (A) The expression of ClyA upon induction with the chemical inducer, L-arabinose. **(B)** The expression of TNF-related apoptosis-inducing ligand under the regulation of the radiation-responsive promoter, *RecA*. **(C)** The expression of ClyA under the control of the acid-sensitive promoter, *adiA*. **(D)** The expression of CTLA-4 and PD-L1-blocking nanobodies under the LuxI/LuxR-type QS system regulation.

harming normal tissues [153]. Anderson et al. [154] developed one of the pioneering systems utilizing AND gates for bacterial therapies. They constructed an AND gate circuit in *E. coli* that produces a bacterial invasin protein encoded by the inv gene from *Yersinia pseudotuberculosis* in response to the input salicylate and Mg^{2+} signals. Invasin induces the bacterial invasion of the non-phagocytic mammalian cells expressing β1-integrin and increases the internalization efficiency of bacteria by the tumor cells. When the promoters of both inputs are activated, bacteria invade HeLa cells [154]. This strategy can be implemented to detect multiple TME signals and obtain improved tumor targeting. As mentioned above, Chien et al. [149] constructed an AND gate circuit in EcN that requires the activation of both hypoxia and lactate-sensitive promoters to express the output genes crucial for bacterial growth [149]. Utilizing this system, the essential genes in the circuit can be replaced with therapeutic agents for cancer, and anti-tumor activity can be achieved.

4.2.2.3 Designing genetic circuits for targeted therapeutic delivery systems

The genetic circuits can be tailored for the therapeutic agents' intracellular or extracellular delivery. Intracellular delivery has the advantages of more accessible access to the target proteins and pathways inside the cell and improved drug molecule stability. To target the intracellular space in tumors, bacteria with intracellular life cycles, such as *S. typhimurium*, are frequently selected as the hosts [136, 155, 156]. However, various engineering strategies enable the use of extracellular bacteria, such as most strains of *E. coli*, for intracellular drug delivery [157].

Gram-negative bacteria with T3SS, such as *S. typhimurium*, can inject the proteins directly into the host cytoplasm [158]. Chabloz et al. [155] improved this system in *S. typhimurium* for the cytosolic delivery of Designed Ankyrin Repeat Proteins and monobodies targeting the RAS pathway. They designed a circuit system in which the pSicA promoter regulated the synthesis of the T3SS effector proteins and the payloads, and the pBAD promoter controlled the hyperinvasive locus A (HilA) protein. This system successfully delivered RAS-inhibiting payloads into cells and effectively blocked RAS signaling in cancer cell lines [155].

To enhance the efficacy of the intracellular delivery of drugs to specific tumors, Park et al. [156] modified the surface of *S. typhimurium* to display a tumor-homing peptide called arginine–glycine–aspartate (RGD) fused with the outer membrane protein A (OmpA) upon induction with L-arabinose. They achieved improved localization to tumors expressing αvβ3 integrins, which are targeted by the RGD peptide (Figure 4.3A) [156].

Another intracellular delivery approach enables the engineered bacteria's internalization into tumor cells. In the research of Critchley-Thorne et al. [157], the extracellular bacteria *E. coli* was programmed to express the invasin protein from *Y. pseudotuberculosis*. In the circuit design, invasin enables the bacterial invasion of tumor cells expressing β1-integrin. The release of the therapeutic payload, ovalbumin, was facilitated by the expression of the pore-forming cytolysin called listeriolysin O (LLO) encoded by the hlyA gene. LLO forms pores on the intracellular vacuole membrane and releases the bacterial content and the plasmids (Figure 4.3B). This strategy paves the way for the use of extracellular bacteria for the intracellular delivery of drugs to tumors [157].

Tumor invasion can also be combined with the lysis of bacteria as an alternative approach for the cytosolic release of drugs. For example, Raman et al. [136] integrated genetic circuits in a strain of *S. enterica*, one of which is an arabinose-inducible circuit expressing the flhDC operon for regulating bacterial motility and tumor invasion. After the tumor invasion facilitated by the expression of the flhDC operon, the second circuit was activated to produce the lysin E gene (lysE) from bacteriophage ΦX1174 to induce the intracellular lysis of *S. enterica* [136].

In addition to intracellular targeting methods, extracellular targeting of tumors is widely used in therapeutic agents. Some of these agents require extracellular delivery because their targeting is mediated through the receptors found on the surface of the tumor cells. Compared to intracellular delivery, extracellular delivery is more straightforward because it does not rely heavily on the physical proximity of bacteria to the tumor cells. Moreover, bacteria can reach higher densities in extracellular space and produce therapeutic agents in greater concentrations. Extracellular targeting also enables inherently extracellular bacteria, such as EcN, to deliver drugs to the tumor sites [134, 146, 159, 160]. However, host selection for extracellular targeting is not limited to extracellular bacteria. Intracellular bacterial species such as *Salmonella* can also be used [161]. In the case of extracellular bacteria, the secretion of the proteins can be challenging. Therefore, these bacteria rely heavily on the signal peptides and secretion tags to release proteins [112]. For instance, Abedi et al. [146] modified EcN to release CTLA-4 and PD-L1 immune checkpoint inhibitors to the extracellular TME with the help of the pelB leader peptide. In the genetic circuitry of this platform, a temperature-sensitive protein, TcI42, represses the production of Bxb1 recombinase at physiological temperatures. Bxb1 is under the regulation of the pL/pR thermally inducible promoters. Upon thermal induction, the repressive activity of TcI42 is inhibited, and Bxb1 expression is observed. Then, Bxb1 acts on the attP and attB sites, flanking the P7 promoter, and inverts the sequence of this promoter in the sense direction. As a result, this promoter drives the expression of CTLA4 and PD-L1 inhibitors with the pelB secretion tag [146].

Although signal peptides and secretion tags play an essential role in enhancing protein secretion in extracellular bacteria, various engineering strategies can be pursued to facilitate the direct delivery of therapeutic agents to tumors without the requirement of these peptides. Reeves et al. [159] established one of the strategies by reprogramming the DH10β strain of *E. coli* to produce a T3SS apparatus from *Shigella flexneri*. This modified strain could directly inject the proteins into the cytoplasm of mammalian cells [159]. In another research, Lynch et al. [162] improved this system further by engineering EcN to express a modified T3SS apparatus from *Shigella* and secrete the proteins to the extracellular space rather than the cytoplasm of the targeted cells. This new system called PROT3EcT (PRObiotic Type III secretion *Escherichia coli* Therapeutic) includes mxi and spa operons encoding for the modified T3SS apparatus. Therapeutic proteins, such as nanobodies targeting TNF-α, were fused with a sequence lacking Ipa operon, which is required to invade the host cells. The activation of these circuits results in the extracellular delivery of antitumor payloads (Figure 4.3C) [162].

Another example of an extracellular targeting system is the EcN strain mentioned above, which Canale et al. [134] modified to improve the antitumor activity. This strain lacks the arginine repressor (ArgR) and produces arginine synthase encoded by the ArgAfbr gene, which is resistant to the negative regulation of L-arginine metabolism. It demonstrated the ability to metabolize ammonia in the TME into L-arginine and release it to the extracellular space to recruit T cells to the tumor sites [134]. Extracellular targeting can also be achieved by harnessing matrix metalloproteinases (MMPs), which are highly abundant in the TME. Therapeutic payloads that contain an MMP target sequence and are expressed on the bacterial cell surface can be delivered to the tumor sites through the cleavage of the target sequence by MMPs [161]. Hyun et al. [161] introduced the MMP cleavage system into the ΔppGpp strain of *S. typhimurium* to produce neoantigens fused with an MMP target sequence. This system facilitated the specific delivery of the neoantigens to the tumor tissues and led to an antitumor effect through T-cell activation (Figure 4.3D).

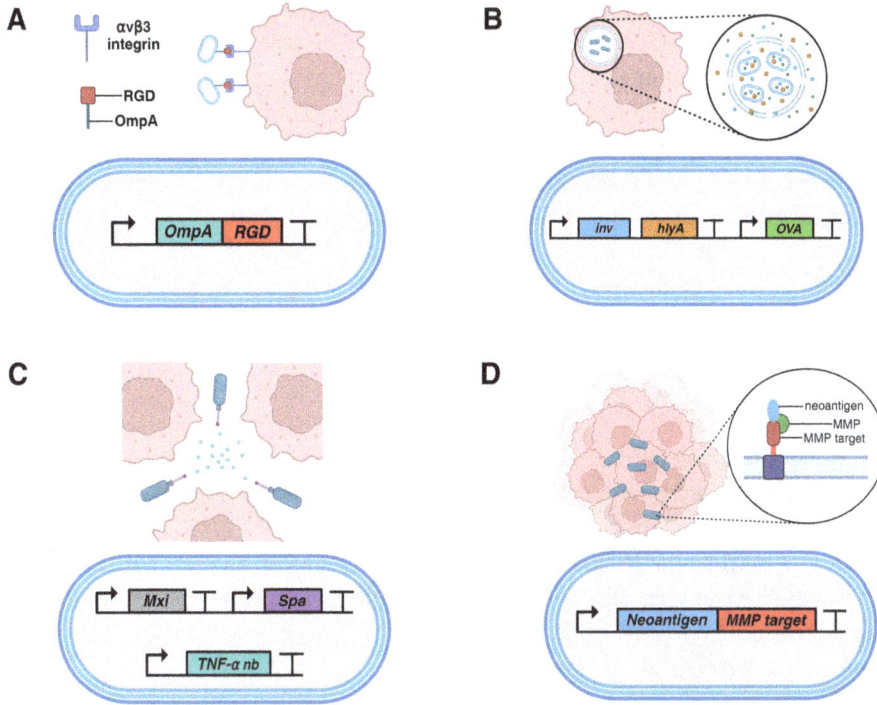

Figure 4.3: Genetic circuits of engineered bacteria for diverse therapeutic delivery approaches: **(A)** enhanced targeting of tumors via surface expression of a tumor-homing peptide, RGD, fused with OmpA; **(B)** bacterial invasion of tumors with *inv* expression and intracellular release of ovalbumin through vacuole membrane pores formed by *hlyA* expression; **(C)** extracellular secretion of TNF-α nanobodies through a modified T3SS apparatus encoded by *mxi* and *spa* operons; and **(D)** local delivery of neoantigens carrying MMP target sequences through tumor-enriched MMP cleavage.

4.3 Bacterial cancer therapy in clinical trials and their regulations for safety

In this section, we will take note of some select case studies, analyze their results, and discuss their success and clinical significance. Once a study passes multiple phases of clinical trials, the treatment can be introduced for commercial use. Considering the case studies discussed, it is probable that shortly we may witness the commercialization of microbe-based therapeutic interventions for cancer.

4.3.1 Studies in clinical trials

Microbial systems are quite adaptive in their approach to cancer management since they can be engineered for direct therapeutic intervention and early cancer detection. Due to its genetic manipulability, proven safety in human applications, and capacity to navigate, endure, and proliferate within the GIT and tumors, EcN stands out as a preferred platform for engineering "smart microbes" [160]. Ongoing clinical trials aim to assess the safety and effectiveness of innovative EcN-based smart microbes, particularly in treating inborn metabolic diseases and cancer. For example, to test if engineered EcN can specifically translocate, colonize, and multiply within liver tumors, EcN was engineered to highly express the lacZ reporter gene and a luminescent lux-CDABE cassette [163]. These probiotics served a dual function of β-galactosidase production and luminescent visualization. Successful colonization was confirmed upon administration, and a tailored substrate, LuGal, was introduced. The LacZ activity within the tumor-resident probiotics converted LuGal to luciferin, a luminescent indicator of liver tumor burden when excreted in the urine. This detection system displayed high sensitivity, requiring only 1 µL of urine for a positive signal within 24 h of probiotic administration. The safety profile of this approach was underscored by the absence of detrimental health effects in mice observed over 12-month post-probiotic ingestion. This study introduces a precise molecular engineering approach and signifies a transformative methodology for noninvasive cancer detection using microbes, promising advancements in diagnostic techniques.

Cancer detection methods using microbes are diverse due to their small size, ease of engineering, numerous species, and various functionalities. One critical characteristic researchers utilize is bacteria's natural competence for DNA uptake. *Acinetobacter baylyi*, a nonpathogenic strain capable of horizontal gene transfer (HGT), was used to develop a bacteria-based cancer detection system [164]. HGT, the transfer of genetic material between nonparental organisms, *A. baylyi*'s transformation and recombination abilities, and ease of use in the lab make it an attractive choice for researchers [165]. The detection system in this study leveraged engineered bacteria to detect specific cell-free DNA sequences from CRC cells, organoids, and tumors. Researchers developed biosensors using *A. baylyi*, which can identify donor DNA from CRC sources both in vitro and in vivo. In vitro experiments demonstrated the biosensors' functionality through coculture assays, while in vivo validation involved delivering sensor bacteria to mice with colorectal tumors. HGT from tumors to biosensors was observed in the murine CRC model, proving the strategy's efficacy in detecting target DNA sequences. This study highlights the potential of bacteria-based diagnostic systems to detect specific mammalian DNA shed from CRC within the gut environment without extensive sample preparation. The long-term objective includes developing an orally administered version capable of reliably detecting target DNA through noninvasive methods, such as stool or blood analysis. This approach represents a transformative advancement in synthetic biology, offering new possibilities for cellular DNA detection and analysis using microbes, with significant implications for future cancer therapies.

One study that reached the trial stage involves a treatment-oriented application of microbes designed for intratumoral injection to activate the innate immune system for cancer treatment [166]. SYNB1891, developed by Synlogic in 2021, harnessed the versatility of nonpathogenic EcN as a living biotherapeutic tailored to treat cancer. SYNB1891 was explicitly designed to target STING activation to phagocytic APCs within tumors. This precise activation triggered complementary innate immune pathways, resulting in potent antitumor immunity and the establishment of immunological memory in murine tumor models. This treatment also exhibited robust activation of human APCs. Advancing into clinical trials, SYNB1891 entered phase I with an intratumoral injection approach for treating solid tumors and lymphoma. The primary aim of this initial trial was to evaluate the safety and tolerability of SYNB1891 when used alone or in conjunction with atezolizumab. Twenty-four individuals received SYNB1891 monotherapy across six groups, while eight participants received combination therapy across two groups. SYNB1891 treatment led to the activation of the STING pathway and upregulation of genes associated with IFN response, chemokine/cytokine production, and T-cell activation, as shown by the analysis of patient tissue samples collected before treatment and 7 days following the third weekly dose. A dose-dependent elevation in serum cytokine levels was also noted [167]. Key findings from the monotherapy cohorts of the trial underscore the safety and tolerability of SYNB1891, demonstrated through intratumoral injection with no observed dose-limiting toxicities or infections. Two patients also displayed evidence of durable, stable disease associated with the upregulation of genes linked to immune activation [168]. Nevertheless, engineered bacteria often struggle to sur-vive in the gut's hostile luminal environment due to competition with native microorganisms and the host's immune response, making persistent colonization difficult [169, 170]. Strategies to enhance bacterial survival and persistence are crucial for the success of such therapies in future clinical applications.

Another exciting approach uses the bacterial strain *Clostridium novyi*-NT, a variant of *Clostridium novyi*. This strain is unique because it lacks the primary toxin gene and is very sensitive to oxygen, providing a new way to fight cancer [171]. When *C. novyi*-NT is combined with traditional chemotherapy or radiation, it is more effective in targeting oxygen-rich cells. This method has improved cure rates in mouse models, especially for large tumors. Initial tests with *C. novyi*-NT had a mortality rate of 10–25%, demonstrating its potential impact [172]. The move from preclinical studies to clinical trials included a phase I trial with a single intravenous infusion of *C. novyi*-NT for solid tumors that did not respond to standard treatments. This strain targeted and destroyed tumors while leaving normal tissue unharmed [173]. Animal tests showed significant tumor shrinkage after just one injection, with about 30% of subjects having no tumor regrowth, which is a considerable achievement. The ongoing phase I study uses a single dose of *C. novyi*-NT spores in patients with treatment-resistant solid tumors to evaluate the safety and early signs of antitumor activity. This highlights the scientific potential of *C. novyi*-NT and its promise to revolutionize cancer treatment for patients who do not respond to conventional therapies.

Another study using the microbe EcN integrates cancer detection and treatment [174]. Researchers showed that orally administered EcN selectively colonized colorectal adenomas in genetically engineered mouse models of CRC and orthotopic CRC. A double-blind, dual-center clinical trial with CRC patients compared those given a placebo to those given EcN for 2 weeks before removing neoplastic and normal colorectal tissue. The trial found that EcN significantly enriched tumor samples in treated patients compared to the normal tissue. The study also explored early CRC intervention strategies by engineering EcN to produce salicylate for lesion detection, locally release GM-CSF, and blocking nanobodies against PD-L1 and CTLA-4 to enhance therapeutic effectiveness. These pathways are crucial immune checkpoints that regulate T-cell responses, and targeting them has revolutionized cancer immunotherapy [175]. Most CRC tumors show microsatellite stability, making them less responsive to immunotherapy [176]. This study addresses this issue by using engineered microbes to boost immune activation. Intratumoral administration of engineered EcN in CRC mouse models significantly reduced adenoma burden. The study demonstrated the feasibility of using engineered EcN for noninvasive adenoma screening through stool and urine samples. Additionally, the research highlighted potential advantages of probiotic platforms over conventional checkpoint therapies, including the possibility of remodeling the TME and improving T-cell infiltration into adenoma cores, thus addressing limitations associated with CRC subtypes.

4.3.2 Regulations of safety and efficacy and concluding remarks

Before LBTs can be translated into real-world applications, stringent safety and efficacy regulations must be followed. The development of these therapies involves extensive preclinical testing to evaluate their safety profile, including the risk of infection, potential toxicity, and unintended effects on the microbiome. Regulatory bodies, such as the FDA, require comprehensive data on the genetic stability of engineered bacteria, their ability to target tumors selectively, and their interactions with the host immune system. Furthermore, clinical trials must demonstrate the therapeutic efficacy of LBTs and long-term safety and tolerability in patients. For instance, testing must include assessments of genetic mutations, environmental release impact, and adverse effects on human health.

Additionally, regulatory oversight includes establishing robust containment and biocontainment measures to prevent accidental release into the environment. This involves testing for HGT potential, ensuring that engineered bacteria do not pass their genetic modifications to other microorganisms in the gut or environment. Post-market surveillance is also essential to monitor the long-term effects of these therapies on patients and the ecosystem. Developers of LBTs must work closely with regulatory bodies throughout the entire process, from initial design to post-approval monitoring, to ensure that these innovative treatments are safe and effective for widespread clinical use. With the various techniques and case studies discussed regarding microbe-based cancer therapeutics, it is clear that this approach marks a significant advancement in medical science. Microorganisms can be engineered to target and destroy cancer cells while leaving healthy cells unharmed, addressing a key challenge in conventional cancer treatments. This method has shown promising results in clinical trials, offering new hope for patients with different types of cancer.

One of the most exciting developments in the medical field is the growing focus on point-of-care devices, which are set to change how we detect and diagnose cancer. This trend can potentially make patient care faster, more accurate, and more accessible. Microbes are playing a pivotal role in developing these devices for cancer detection. Moreover, the advantages of a microorganism-centered approach to cancer treatment go beyond detection and diagnostics. This approach is a stepping stone toward personalized medicine in cancer. Personalized medicine tailors medical treatment to an individual's unique genetic, environmental, and lifestyle factors. This approach recognizes that each person is different and that a one-size-fits-all approach to medicine is often insufficient. For instance, the multi-faceted nature of triple-negative breast cancer, characterized by its diverse molecular profiles and variable treatment responses in different patients, makes it challenging to treat using a single method. By using advanced genomics, proteomics, and other omics fields, personalized medicine seeks to identify specific molecular mechanisms driving disease in each patient. Microbes have long been an unexplored domain, but research shows their vast potential as tools for personalized medicine, and there is now increasing interest among researchers in tapping into this potential.

In personalized medicine, microbes can selectively target cancer cells, deliver drugs directly to tumors, and even stimulate the immune system to attack cancer cells. Researchers can genetically engineer bacteria to produce drugs or carry drug-loaded particles to tumor sites. These bacteria can colonize tumor sites, as in the case of EcN discussed above. The bacteria release the drugs at the tumor site, reducing systemic toxicity and increasing drug concentration. Microbes can also stimulate the immune system to recognize and attack cancer cells. Bacteria that directly activate immune cells (DCs, T cells, and NK cells) may also produce immunomodulatory compounds, enhancing the patient's immune response. Microbes can produce specific metabolites or biomarkers in response to cancer cell interactions. These biomarkers can be used for early detection, monitoring, and personalized treatment. They can also indicate treatment response and detect recurrences. These mechanisms show promise in developing targeted, effective, and precision treatments for cancer patients, potentially improving patient outcomes and quality of life.

In conclusion, the development of microbe-based cancer therapeutics has led to significant advancements in oncology, enabling the creation of targeted and innovative treatment strategies. By using the unique properties of microbes, such as their ability to colonize tumor sites and interact with the immune system selectively, researchers have developed new approaches to cancer treatment. These include microbe-based drug delivery systems, which allow targeted and efficient drug release, and microbe-based immunotherapies, which stimulate the immune system to recognize and attack cancer cells. In addition, microbes can produce specific biomarkers, allowing for early detection and monitoring of cancer progression. As seen in case studies, improved treatment outcomes and enhanced patient survival rates were observed through these approaches. Furthermore, using microbes as therapeutic agents has shown promise in overcoming traditional chemotherapy resistance and minimizing adverse effects. Researchers continue to investigate the complex interactions between microbes and cancer cells to understand the underlying mechanisms and optimize treatment strategies. These ongoing research efforts aim to improve the design and delivery of microbe-based therapeutics, enhancing their efficacy and safety. Researchers are also exploring the potential for microbes to serve as vectors for gene therapy and tools for cancer diagnosis and monitoring. By continuing to push the boundaries of this approach, there is hope for making a meaningful impact on patient care and outcomes. Delving deeper into the relationship between microbes and cancer cells will reveal new ways to use these tiny agents for therapeutic gain. As this relationship continues to be understood, a comprehensive understanding of cancer biology and treatment strategies may be achieved. Continued research and collaboration in microbial therapeutics will shape the future of cancer patient care.

References

[1] Ursell LK, Metcalf JL, Parfrey LW, Knight R. Defining the human microbiome. Nutr Rev 2012;70:S38–44. https://doi.org/10.1111/j.1753-4887.2012.00493.x.

[2] Lynch SV, Pedersen O. The human intestinal microbiome in health and disease. N Engl J Med 2016;375:2369–79. https://doi.org/10.1056/NEJMra1600266.

[3] Kogut MH, Lee A, Santin E. Microbiome and pathogen interaction with the immune system. Poult Sci 2020;99:1906–13. https://doi.org/10.1016/j.psj.2019.12.011.

[4] Cho I, Blaser MJ. The human microbiome: at the interface of health and disease. Nat Rev Genet 2012;13:260–70. https://doi.org/10.1038/nrg3182.

[5] Schwabe RF, Jobin C. The microbiome and cancer. Nat Rev Cancer 2013;13:800–12. https://doi.org/10.1038/nrc3610.

[6]   Muhammad AY, Amonov M, Baig AA, Alvi FJ. Gut microbiome: an intersection between human genome, diet, and epigenetics. Advanced Gut & Microbiome Research 2024;2024:e6707728. https://doi.org/10.1155/2024/6707728.

[7]   Sender R, Fuchs S, Milo R. Revised estimates for the number of human and bacteria cells in the body. PLoS Biol 2016;14:e1002533. https://doi.org/10.1371/journal.pbio.1002533.

[8]   Gill SR, Pop M, DeBoy RT, Eckburg PB, Turnbaugh PJ, Samuel BS, et al. Metagenomic analysis of the human distal gut microbiome. Science 2006;312:1355–59. https://doi.org/10.1126/science.1124234.

[9]   Bäckhed F, Ley RE, Sonnenburg JL, Peterson DA, Gordon JI. Host-bacterial mutualism in the human intestine. Science 2005;307:1915–20. https://doi.org/10.1126/science.1104816.

[10]   Hooper LV, Gordon JI. Commensal host-bacterial relationships in the gut. Science 2001;292:1115–18. https://doi.org/10.1126/science.1058709.

[11]   Mishra Y, Ranjan A, Mishra V, Chattaraj A, Aljabali AAA, El-Tanani M, et al. The role of the gut microbiome in gastrointestinal cancers. Cell Signal 2024;115:111013. https://doi.org/10.1016/j.cellsig.2023.111013.

[12]   Malla MA, Dubey A, Kumar A, Yadav S, Hashem A, Abd_Allah EF. Exploring the human microbiome: the potential future role of next-generation sequencing in disease diagnosis and treatment. Front Immunol 2019;9. https://doi.org/10.3389/fimmu.2018.02868.

[13]   Wensel CR, Pluznick JL, Salzberg SL, Sears CL. Next-generation sequencing: insights to advance clinical investigations of the microbiome. J Clin Invest 2022;132. https://doi.org/10.1172/JCI154944.

[14]   Yi X, Lu H, Liu X, He J, Li B, Wang Z, et al. Unravelling the enigma of the human microbiome: evolution and selection of sequencing technologies. Microb Biotechnol 2024;17:e14364. https://doi.org/10.1111/1751-7915.14364.

[15]   Turnbaugh PJ, Ley RE, Hamady M, Fraser-Liggett CM, Knight R, Gordon JI. The human microbiome project. Nature 2007;449:804–10. https://doi.org/10.1038/nature06244.

[16]   Qin J, Li R, Raes J, Arumugam M, Burgdorf KS, Manichanh C, et al. A human gut microbial gene catalogue established by metagenomic sequencing. Nature 2010;464:59–65. https://doi.org/10.1038/nature08821.

[17]   Turnbaugh PJ, Ley RE, Hamady M, Fraser-Liggett C, Knight R, Gordon JI. The human microbiome project: exploring the microbial part of ourselves in a changing world. Nature 2007;449:804–10. https://doi.org/10.1038/nature06244.

[18]   Staudinger T, Pipal A, Redl B. Molecular analysis of the prevalent microbiota of human male and female forehead skin compared to forearm skin and the influence of make-up. J Appl Microbiol 2011;110:1381–89. https://doi.org/10.1111/j.1365-2672.2011.04991.x.

[19]   Pei Z, Bini EJ, Yang L, Zhou M, Francois F, Blaser MJ. Bacterial biota in the human distal esophagus. Proc Natl Acad Sci 2004;101:4250–55. https://doi.org/10.1073/pnas.0306398101.

[20]   Bik EM, Eckburg PB, Gill SR, Nelson KE, Purdom EA, Francois F, et al. Molecular analysis of the bacterial microbiota in the human stomach. Proc Natl Acad Sci 2006;103:732–37. https://doi.org/10.1073/pnas.0506655103.

[21]   Eckburg PB, Bik EM, Bernstein CN, Purdom E, Dethlefsen L, Sargent M, et al. Diversity of the human intestinal microbial flora. Science 2005;308:1635–38. https://doi.org/10.1126/science.1110591.

[22]   Ley RE, Turnbaugh PJ, Klein S, Gordon JI. Microbial ecology: human gut microbes associated with obesity. Nature 2006;444:1022–23. https://doi.org/10.1038/4441022a.

[23]   Hyman RW, Fukushima M, Diamond L, Kumm J, Giudice LC, Davis RW. Microbes on the human vaginal epithelium. Proc Natl Acad Sci U S A 2005;102:7952–57. https://doi.org/10.1073/pnas.0503236102.

[24]   Kinross JM, Darzi AW, Nicholson JK. Gut microbiome-host interactions in health and disease. Genome Med 2011;3:14. https://doi.org/10.1186/gm228.

[25]   Lamendella R, VerBerkmoes N, Jansson JK. 'Omics' of the mammalian gut – new insights into function. Curr Opin Biotech 2012;23:491–500. https://doi.org/10.1016/j.copbio.2012.01.016.

[26] Whon TW, Shin N-R, Kim JY, Roh SW. Omics in gut microbiome analysis. J Microbiol 2021;59:292–97. https://doi.org/10.1007/s12275-021-1004-0.

[27] Tang WHW, Kitai T, Hazen SL. Gut microbiota in cardiovascular health and disease. Circ Res 2017;120:1183–96. https://doi.org/10.1161/CIRCRESAHA.117.309715.

[28] Tang WHW, Li DY, Hazen SL. Dietary metabolism, the gut microbiome, and heart failure. Nat Rev Cardiol 2019;16:137–54. https://doi.org/10.1038/s41569-018-0108-7.

[29] Glassner KL, Abraham BP, Quigley EMM. The microbiome and inflammatory bowel disease. J Allergy Clin Immunol 2020;145:16–27. https://doi.org/10.1016/j.jaci.2019.11.003.

[30] Ning L, Zhou Y-L, Sun H, Zhang Y, Shen C, Wang Z, et al. Microbiome and metabolome features in inflammatory bowel disease via multi-omics integration analyses across cohorts. Nat Commun 2023;14:7135. https://doi.org/10.1038/s41467-023-42788-0.

[31] Zhao Q, Chen Y, Huang W, Zhou H, Zhang W. Drug-microbiota interactions: an emerging priority for precision medicine. Sig Transduct Target Ther 2023;8:1–27. https://doi.org/10.1038/s41392-023-01619-w.

[32] Rosell-Mases E, Santiago A, Corral-Pujol M, Yáñez F, Varela E, Egia-Mendikute L, et al. Mutual modulation of gut microbiota and the immune system in type 1 diabetes models. Nat Commun 2023;14:7770. https://doi.org/10.1038/s41467-023-43652-x.

[33] Zheng D, Liwinski T, Elinav E. Interaction between microbiota and immunity in health and disease. Cell Res 2020;30:492–506. https://doi.org/10.1038/s41422-020-0332-7.

[34] Liu Y, Wang J, Wu C. Modulation of gut microbiota and immune system by probiotics, pre-biotics, and post-biotics. Front Nutr 2021;8:634897. https://doi.org/10.3389/fnut.2021.634897.

[35] Song X, Sun X, Oh SF, Wu M, Zhang Y, Zheng W, et al. Microbial bile acid metabolites modulate gut RORγ+ regulatory T cell homeostasis. Nature 2020;577:410–15. https://doi.org/10.1038/s41586-019-1865-0.

[36] Zheng Z, Hu Y, Tang J, Xu W, Zhu W, Zhang W. The implication of gut microbiota in recovery from gastrointestinal surgery. Front Cell Infect Microbiol 2023;13:1110787. https://doi.org/10.3389/fcimb.2023.1110787.

[37] Busch W. Aus der Sitzung der medicinischen Section vom 13 November 1867. Berliner Klinische Wochenschrift 1868;5:137.

[38] Fehleisen F. Ueber die Züchtung der Erysipelkokken auf künstlichem Nährboden und ihre Übertragbarkeit auf den Menschen. Dtsch Med Wochenschr 1882;8:553–54. https://doi.org/10.1055/s-0029-1196806.

[39] Starnes CO. Coley's toxins in perspective. Nature 1992;357:11–12. https://doi.org/10.1038/357011a0.

[40] Sholl J, Sepich-Poore GD, Knight R, Pradeu T. Redrawing therapeutic boundaries: microbiota and cancer. Trends Cancer 2022;8:87–97. https://doi.org/10.1016/j.trecan.2021.10.008.

[41] Hanahan D, Weinberg RA. Hallmarks of cancer: the next generation. Cell 2011;144:646–74. https://doi.org/10.1016/j.cell.2011.02.013.

[42] Sepich-Poore GD, Zitvogel L, Straussman R, Hasty J, Wargo JA, Knight R. The microbiome and human cancer. Science 2021;371:eabc4552. https://doi.org/10.1126/science.abc4552.

[43] Cullin N, Azevedo Antunes C, Straussman R, Stein-Thoeringer CK, Elinav E. Microbiome and cancer. Cancer Cell 2021;39:1317–41. https://doi.org/10.1016/j.ccell.2021.08.006.

[44] Helmink BA, Khan MAW, Hermann A, Gopalakrishnan V, Wargo JA. The microbiome, cancer, and cancer therapy. Nat Med 2019;25:377–88. https://doi.org/10.1038/s41591-019-0377-7.

[45] Bleich RM, Arthur JC. Microbiome and the Hallmarks of Cancer. In: Sun J, editor. Inflammation, Infection, and Microbiome in Cancers: Evidence, Mechanisms, and Implications, Cham: Springer International Publishing; 2021, pp. 1–26. https://doi.org/10.1007/978-3-030-67951-4_1.

[46] Xuan C, Shamonki JM, Chung A, Dinome ML, Chung M, Sieling PA, et al. Microbial dysbiosis is associated with human breast cancer. PLoS One 2014;9:e83744. https://doi.org/10.1371/journal.pone.0083744.

[47] Pereira-Marques J, Ferreira RM, Pinto-Ribeiro I, Figueiredo C. Helicobacter pylori infection, the gastric microbiome and gastric cancer. Adv Exp Med Biol 2019;1149:195–210. https://doi.org/10.1007/5584_2019_366.

[48] Mager LF, Burkhard R, Pett N, Cooke NCA, Brown K, Ramay H, et al. Microbiome-derived inosine modulates response to checkpoint inhibitor immunotherapy. Science 2020;369:1481–89. https://doi.org/10.1126/science.abc3421.

[49] Belkaid Y, Naik S. Compartmentalized and systemic control of tissue immunity by commensals. Nat Immunol 2013;14:646–53. https://doi.org/10.1038/ni.2604.

[50] Fung TC, Olson CA, Hsiao EY. Interactions between the microbiota, immune and nervous systems in health and disease. Nat Neurosci 2017;20:145–55. https://doi.org/10.1038/nn.4476.

[51] Zitvogel L, Ma Y, Raoult D, Kroemer G, Gajewski TF. The microbiome in cancer immunotherapy: diagnostic tools and therapeutic strategies. Science 2018;359:1366–70. https://doi.org/10.1126/science.aar6918.

[52] Nejman D, Livyatan I, Fuks G, Gavert N, Zwang Y, Geller LT, et al. The human tumor microbiome is composed of tumor type–specific intracellular bacteria. Science 2020;368:973–80. https://doi.org/10.1126/science.aay9189.

[53] Pushalkar S, Hundeyin M, Daley D, Zambirinis CP, Kurz E, Mishra A, et al. The pancreatic cancer microbiome promotes oncogenesis by induction of innate and adaptive immune suppression. Cancer Discov 2018;8:403–16. https://doi.org/10.1158/2159-8290.CD-17-1134.

[54] Jin C, Lagoudas GK, Zhao C, Bullman S, Bhutkar A, Hu B, et al. Commensal microbiota promote lung cancer development via γδ T cells. Cell 2019;176:998–1013, e16. https://doi.org/10.1016/j.cell.2018.12.040.

[55] Bullman S, Pedamallu CS, Sicinska E, Clancy TE, Zhang X, Cai D, et al. Analysis of Fusobacterium persistence and antibiotic response in colorectal cancer. Science 2017;358:1443–48. https://doi.org/10.1126/science.aal5240.

[56] Schwabe RF, Jobin C. The microbiome and cancer. Nat Rev Cancer 2013;13:800–12. https://doi.org/10.1038/nrc3610.

[57] Foster KR, Schluter J, Coyte KZ, Rakoff-Nahoum S. The evolution of the host microbiome as an ecosystem on a leash. Nature 2017;548:43–51. https://doi.org/10.1038/nature23292.

[58] Almeida A, Mitchell AL, Boland M, Forster SC, Gloor GB, Tarkowska A, et al. A new genomic blueprint of the human gut microbiota. Nature 2019;568:499–504. https://doi.org/10.1038/s41586-019-0965-1.

[59] Iarc. Iarc Monographs on the Identification of Carcinogenic Hazards to Humans 2009.

[60] Ishaq S, Nunn L. Helicobacter pylori and gastric cancer: a state of the art review. Gastroenterol Hepatol Bed Bench 2015;8:S6–14.

[61] Song X, Xin N, Wang W, Zhao C. Wnt/β-catenin, an oncogenic pathway targeted by H. pylori in gastric carcinogenesis. Oncotarget 2015;6:35579–88.

[62] Zuo W, Yang H, Li N, Ouyang Y, Xu X, Hong J. Helicobacter pylori infection activates Wnt/β-catenin pathway to promote the occurrence of gastritis by upregulating ASCL1 and AQP5. Cell Death Discov 2022;8:1–10. https://doi.org/10.1038/s41420-022-01026-0.

[63] Franco AT, Israel DA, Washington MK, Krishna U, Fox JG, Rogers AB, et al. Activation of β-catenin by carcinogenic Helicobacter pylori. Proc Natl Acad Sci 2005;102:10646–51. https://doi.org/10.1073/pnas.0504927102.

[64] Hamway Y, Taxauer K, Moonens K, Neumeyer V, Fischer W, Schmitt V, et al. Cysteine residues in helicobacter pylori adhesin HopQ are required for CEACAM–HopQ interaction and subsequent CagA translocation. Microorganisms 2020;8:465. https://doi.org/10.3390/microorganisms8040465.

[65] Javaheri A, Kruse T, Moonens K, Mejías-Luque R, Debraekeleer A, Asche CI, et al. Helicobacter pylori adhesin HopQ engages in a virulence-enhancing interaction with human CEACAMs. Nat Microbiol 2016;2:1–13. https://doi.org/10.1038/nmicrobiol.2016.189.

[66] Odenbreit S, Püls J, Sedlmaier B, Gerland E, Fischer W, Haas R. Translocation of Helicobacter pylori CagA into gastric epithelial cells by type IV secretion. Science 2000;287:1497–500. https://doi.org/10.1126/science.287.5457.1497.

[67] Murata-Kamiya N, Kurashima Y, Teishikata Y, Yamahashi Y, Saito Y, Higashi H, et al. Helicobacter pylori CagA interacts with E-cadherin and deregulates the β-catenin signal that promotes intestinal transdifferentiation in gastric epithelial cells. Oncogene 2007;26:4617–26. https://doi.org/10.1038/sj.onc.1210251.

[68] Silva-García O, Valdez-Alarcón JJ, Baizabal-Aguirre VM. Wnt/β-catenin signaling as a molecular target by pathogenic bacteria. Front Immunol 2019;10:2135. https://doi.org/10.3389/fimmu.2019.02135.

[69] Hatakeyama M. Structure and function of Helicobacter pylori CagA, the first-identified bacterial protein involved in human cancer. Proc Jpn Acad Ser B Phys Biol Sci 2017;93:196–219. https://doi.org/10.2183/pjab.93.013.

[70] Pleguezuelos-Manzano C, Puschhof J, Rosendahl Huber A, van Hoeck A, Wood HM, Nomburg J, et al. Mutational signature in colorectal cancer caused by genotoxic pks+ E. coli. Nature 2020;580:269–73. https://doi.org/10.1038/s41586-020-2080-8.

[71] Wilson MR, Jiang Y, Villalta PW, Stornetta A, Boudreau PD, Carrá A, et al. The human gut bacterial genotoxin colibactin alkylates DNA. Science 2019;363:eaar7785. https://doi.org/10.1126/science.aar7785.

[72] Dejea CM, Fathi P, Craig JM, Boleij A, Taddese R, Geis AL, et al. Patients with familial adenomatous polyposis harbor colonic biofilms containing tumorigenic bacteria. Science 2018;359:592–97. https://doi.org/10.1126/science.aah3648.

[73] Barrett M, Hand CK, Shanahan F, Murphy T, O'Toole PW. Mutagenesis by microbe: the role of the microbiota in shaping the cancer genome. Trends in Cancer 2020;6:277–87. https://doi.org/10.1016/j.trecan.2020.01.019.

[74] Rubinstein MR, Wang X, Liu W, Hao Y, Cai G, Han YW. Fusobacterium nucleatum promotes colorectal carcinogenesis by modulating E-cadherin/β-catenin signaling via its FadA adhesin. Cell Host Microbe 2013;14:195–206. https://doi.org/10.1016/j.chom.2013.07.012.

[75] Kadosh E, Snir-Alkalay I, Venkatachalam A, May S, Lasry A, Elyada E, et al. The gut microbiome switches mutant p53 from tumour-suppressive to oncogenic. Nature 2020;586:133–38. https://doi.org/10.1038/s41586-020-2541-0.

[76] Bossuet-Greif N, Vignard J, Taieb F, Mirey G, Dubois D, Petit C, et al. The colibactin genotoxin generates DNA interstrand cross-links in infected cells. mBio 2018;9:e02393–17. https://doi.org/10.1128/mBio.02393-17.

[77] Haghi F, Goli E, Mirzaei B, Zeighami H. The association between fecal enterotoxigenic B. fragilis with colorectal cancer. BMC Cancer 2019;19:879. https://doi.org/10.1186/s12885-019-6115-1.

[78] Zepeda-Rivera M, Minot SS, Bouzek H, Wu H, Blanco-Míguez A, Manghi P, et al. A distinct Fusobacterium nucleatum clade dominates the colorectal cancer niche. Nature 2024;628:424–32. https://doi.org/10.1038/s41586-024-07182-w.

[79] Ou S, Wang H, Tao Y, Luo K, Ye J, Ran S, et al. Fusobacterium nucleatum and colorectal cancer: from phenomenon to mechanism. Front Cell Infect Microbiol 2022;12:1020583. https://doi.org/10.3389/fcimb.2022.1020583.

[80] Rubinstein MR, Baik JE, Lagana SM, Han RP, Raab WJ, Sahoo D, et al. Fusobacterium nucleatum promotes colorectal cancer by inducing Wnt/β-catenin modulator annexin A1. EMBO Reports 2019;20:e47638. https://doi.org/10.15252/embr.201847638.

[81] Wu S, Ye Z, Liu X, Zhao Y, Xia Y, Steiner A, et al. Salmonella typhimurium infection increases p53 acetylation in intestinal epithelial cells. Am J Physiol Gastrointest Liver Physiol 2010;298:G784–94. https://doi.org/10.1152/ajpgi.00526.2009.

[82] Lu R, Bosland M, Xia Y, Zhang Y-G, Kato I, Sun J. Presence of Salmonella AvrA in colorectal tumor and its precursor lesions in mouse intestine and human specimens. Oncotarget 2017;8:55104–15. https://doi.org/10.18632/oncotarget.19052.

[83] Levy M, Kolodziejczyk AA, Thaiss CA, Elinav E. Dysbiosis and the immune system. Nat Rev Immunol 2017;17:219–32. https://doi.org/10.1038/nri.2017.7.

[84] Nakatsu G, Li X, Zhou H, Sheng J, Wong SH, Wu WKK, et al. Gut mucosal microbiome across stages of colorectal carcinogenesis. Nat Commun 2015;6:8727. https://doi.org/10.1038/ncomms9727.

[85] Wirbel J, Pyl PT, Kartal E, Zych K, Kashani A, Milanese A, et al. Meta-analysis of fecal metagenomes reveals global microbial signatures that are specific for colorectal cancer. Nat Med 2019;25:679–89. https://doi.org/10.1038/s41591-019-0406-6.

[86] Mima K, Sukawa Y, Nishihara R, Qian ZR, Yamauchi M, Inamura K, et al. Fusobacterium nucleatum and T cells in colorectal carcinoma. JAMA Oncol 2015;1:653–61. https://doi.org/10.1001/jamaoncol. 2015.1377.

[87] Riquelme E, Zhang Y, Zhang L, Montiel M, Zoltan M, Dong W, et al. Tumor microbiome diversity and composition influence pancreatic cancer outcomes. Cell 2019;178:795–806, e12. https://doi.org/ 10.1016/j.cell.2019.07.008.

[88] Hamada T, Zhang X, Mima K, Bullman S, Sukawa Y, Nowak JA, et al. Fusobacterium nucleatum in colorectal cancer relates to immune response differentially by tumor microsatellite instability status. Cancer Immunol Res 2018;6:1327–36. https://doi.org/10.1158/2326-6066.CIR-18-0174.

[89] Parhi L, Alon-Maimon T, Sol A, Nejman D, Shhadeh A, Fainsod-Levi T, et al. Breast cancer colonization by Fusobacterium nucleatum accelerates tumor growth and metastatic progression. Nat Commun 2020;11:3259. https://doi.org/10.1038/s41467-020-16967-2.

[90] Yu T, Guo F, Yu Y, Sun T, Ma D, Han J, et al. Fusobacterium nucleatum promotes chemoresistance to colorectal cancer by modulating autophagy. Cell 2017;170:548–63, e16. https://doi.org/10.1016/j.cell. 2017.07.008.

[91] Zhou S, Gravekamp C, Bermudes D, Liu K. Tumour-targeting bacteria engineered to fight cancer. Nat Rev Cancer 2018;18:727–43. https://doi.org/10.1038/s41568-018-0070-z.

[92] Chu H, Mazmanian SK. Innate immune recognition of the microbiota promotes host-microbial symbiosis. Nat Immunol 2013;14:668–75. https://doi.org/10.1038/ni.2635.

[93] Qiu Q, Lin Y, Ma Y, Li X, Liang J, Chen Z, et al. Exploring the emerging role of the gut microbiota and tumor microenvironment in cancer immunotherapy. Front Immunol 2021;11. https://doi.org/ 10.3389/fimmu.2020.612202.

[94] Luchner M, Reinke S, Milicic A. TLR agonists as vaccine adjuvants targeting cancer and infectious diseases. Pharmaceutics 2021;13:142. https://doi.org/10.3390/pharmaceutics13020142.

[95] Shen Y, Giardino Torchia ML, Lawson GW, Karp CL, Ashwell JD, Mazmanian SK. Outer membrane vesicles of a human commensal mediate immune regulation and disease protection. Cell Host Microbe 2012;12:509–20. https://doi.org/10.1016/j.chom.2012.08.004.

[96] Liu W, Chen Q, Wu S, Xia X, Wu A, Cui F, et al. Radioprotector WR-2721 and mitigating peptidoglycan synergistically promote mouse survival through the amelioration of intestinal and bone marrow damage. J Radiat Res 2015;56:278–86. https://doi.org/10.1093/jrr/rru100.

[97] Burdelya LG, Krivokrysenko VI, Tallant TC, Strom E, Gleiberman AS, Gupta D, et al. An agonist of toll-like receptor 5 has radioprotective activity in mouse and primate models. Science 2008;320:226–30. https://doi.org/10.1126/science.1154986.

[98] Fischer JC, Bscheider M, Eisenkolb G, Lin C-C, Wintges A, Otten V, et al. RIG-I/MAVS and STING signaling promote gut integrity during irradiation- and immune-mediated tissue injury. Sci Transl Med 2017;9:eaag2513. https://doi.org/10.1126/scitranslmed.aag2513.

[99] Viaud S, Saccheri F, Mignot G, Yamazaki T, Daillère R, Hannani D, et al. The intestinal microbiota modulates the anticancer immune effects of cyclophosphamide. Science 2013;342:971–76. https://doi.org/10.1126/science.1240537.

[100] Daillère R, Vétizou M, Waldschmitt N, Yamazaki T, Isnard C, Poirier-Colame V, et al. *Enterococcus hirae* and *Barnesiella intestinihominis* facilitate cyclophosphamide-induced therapeutic immunomodulatory effects. Immunity 2016;45:931–43. https://doi.org/10.1016/j.immuni.2016.09.009.

[101] Vétizou M, Pitt JM, Daillère R, Lepage P, Waldschmitt N, Flament C, et al. Anticancer immunotherapy by CTLA-4 blockade relies on the gut microbiota. Science 2015;350:1079–84. https://doi.org/10.1126/science.aad1329.

[102] Routy B, Le Chatelier E, Derosa L, Duong CPM, Alou MT, Daillère R, et al. Gut microbiome influences efficacy of PD-1-based immunotherapy against epithelial tumors. Science 2018;359:91–97. https://doi.org/10.1126/science.aan3706.

[103] Uribe-Herranz M, Rafail S, Beghi S, Gil-de-Gómez L, Verginadis I, Bittinger K, et al. Gut microbiota modulate dendritic cell antigen presentation and radiotherapy-induced antitumor immune response. J Clin Invest 2020;130:466–79. https://doi.org/10.1172/JCI124332.

[104] Ansaldo E, Slayden LC, Ching KL, Koch MA, Wolf NK, Plichta DR, et al. Akkermansia muciniphila induces intestinal adaptive immune responses during homeostasis. Science 2019;364:1179–84. https://doi.org/10.1126/science.aaw7479.

[105] Wang Y, Wiesnoski DH, Helmink BA, Gopalakrishnan V, Choi K, DuPont HL, et al. Fecal microbiota transplantation for refractory immune checkpoint inhibitor-associated colitis. Nat Med 2018;24:1804–08. https://doi.org/10.1038/s41591-018-0238-9.

[106] van Nood E, Vrieze A, Nieuwdorp M, Fuentes S, Zoetendal EG, de Vos WM, et al. Duodenal infusion of donor feces for recurrent Clostridium difficile. N Engl J Med 2013;368:407–15. https://doi.org/10.1056/NEJMoa1205037.

[107] Gopalakrishnan V, Spencer CN, Nezi L, Reuben A, Andrews MC, Karpinets TV, et al. Gut microbiome modulates response to anti-PD-1 immunotherapy in melanoma patients. Science 2018;359:97–103. https://doi.org/10.1126/science.aan4236.

[108] Routy B, Le Chatelier E, Derosa L, Duong CPM, Alou MT, Daillère R, et al. Gut microbiome influences efficacy of PD-1–based immunotherapy against epithelial tumors. Science 2018;359:91–97. https://doi.org/10.1126/science.aan3706.

[109] Baruch EN, Youngster I, Ben-Betzalel G, Ortenberg R, Lahat A, Katz L, et al. Fecal microbiota transplant promotes response in immunotherapy-refractory melanoma patients. Science 2021;371:602–09. https://doi.org/10.1126/science.abb5920.

[110] Steck SE, Murphy EA. Dietary patterns and cancer risk. Nat Rev Cancer 2020;20:125–38. https://doi.org/10.1038/s41568-019-0227-4.

[111] Forbes NS. Engineering the perfect (bacterial) cancer therapy. Nat Rev Cancer 2010;10:785–94. https://doi.org/10.1038/nrc2934.

[112] Gurbatri CR, Arpaia N, Danino T. Engineering bacteria as interactive cancer therapies. Science 2022;378:858–64. https://doi.org/10.1126/science.add9667.

[113] Charbonneau MR, Isabella VM, Li N, Kurtz CB. Developing a new class of engineered live bacterial therapeutics to treat human diseases. Nat Commun 2020;11:1738. https://doi.org/10.1038/s41467-020-15508-1.

[114] Turnbaugh PJ, Ley RE, Hamady M, Fraser-Liggett CM, Knight R, Gordon JI. The human microbiome project. Nature 2007;449:804–10. https://doi.org/10.1038/nature06244.

[115] Raman V, Deshpande CP, Khanduja S, Howell LM, Van Dessel N, Forbes NS. Build-a-bug workshop: using microbial-host interactions and synthetic biology tools to create cancer therapies. Cell Host & Microbe 2023;31:1574–92. https://doi.org/10.1016/j.chom.2023.09.006.

[116] Fu A, Yao B, Dong T, Chen Y, Yao J, Liu Y, et al. Tumor-resident intracellular microbiota promotes metastatic colonization in breast cancer. Cell 2022;185:1356–72, e26. https://doi.org/10.1016/j.cell.2022.02.027.

[117] Hiyoshi H, English BC, Diaz-Ochoa VE, Wangdi T, Zhang LF, Sakaguchi M, et al. Virulence factors perforate the pathogen-containing vacuole to signal efferocytosis. Cell Host & Microbe 2022;30:163–70, e6. https://doi.org/10.1016/j.chom.2021.12.001.

[118] Tan W, Duong MT-Q, Zuo C, Qin Y, Zhang Y, Guo Y, et al. Targeting of pancreatic cancer cells and stromal cells using engineered oncolytic *Salmonella typhimurium*. Mol Ther 2022;30:662–71. https://doi.org/10.1016/j.ymthe.2021.08.023.

[119] Knodler LA, Vallance BA, Celli J, Winfree S, Hansen B, Montero M, et al. Dissemination of invasive Salmonella via bacterial-induced extrusion of mucosal epithelia. Proc Natl Acad Sci 2010;107:17733–38. https://doi.org/10.1073/pnas.1006098107.

[120] Gut inflammation provides a respiratory electron acceptor for Salmonella | Nature n.d. https://www.nature.com/articles/nature09415 (accessed June 14, 2024).

[121] Palmer C, Bik EM, DiGiulio DB, Relman DA, Brown PO. Development of the human infant intestinal microbiota. PLOS Biology 2007;5:e177. https://doi.org/10.1371/journal.pbio.0050177.

[122] The intestinal barrier: a fundamental role in health and disease. Expert Rev Gastroenterol Hepatol Vol 11, No 9 – Get Access n.d. https://www.tandfonline.com/doi/full/10.1080/17474124.2017.1343143 (accessed June 14, 2024).

[123] Maltby R, Leatham-Jensen MP, Gibson T, Cohen PS, Conway T. Nutritional basis for colonization resistance by human commensal Escherichia coli strains HS and Nissle 1917 against E. coli O157:H7 in the mouse intestine. PLOS ONE 2013;8:e53957. https://doi.org/10.1371/journal.pone.0053957.

[124] Wang C, Shen Y, Ma Y. Bifidobacterium infantis-mediated herpes simplex virus-TK/ganciclovir treatment inhibits cancer metastasis in mouse model. Int J Mol Sci 2023;24:11721. https://doi.org/10.3390/ijms241411721.

[125] Xu X, Zhang M, Liu X, Chai M, Diao L, Ma L, et al. Probiotics formulation and cancer nanovaccines show synergistic effect in immunotherapy and prevention of colon cancer. iScience 2023;26:107167. https://doi.org/10.1016/j.isci.2023.107167.

[126] Abo-Zaid GA, Kenawy AM, El-Deeb NM, Al-Madboly LA. Improvement and enhancement of oligosaccharide production from Lactobacillus acidophilus using statistical experimental designs and its inhibitory effect on colon cancer. Microb Cell Fact 2023;22:148. https://doi.org/10.1186/s12934-023-02153-8.

[127] Hatami S, Yavarmanesh M, Sankian M. Comparison of the effects of probiotic strains (Lactobacillus gasseri, Lactiplantibacillus plantarum, Lactobacillus acidophilus, and Limosilactobacillus fermentum) isolated from human and food products on the immune response of CT26 tumor–bearing mice. Braz J Microbiol 2023;54:2047–62. https://doi.org/10.1007/s42770-023-01060-9.

[128] Nguyen D-H, Chong A, Hong Y, Min -J-J. Bioengineering of bacteria for cancer immunotherapy. Nat Commun 2023;14:3553. https://doi.org/10.1038/s41467-023-39224-8.

[129] Feng Z, Wang Y, Xu H, Guo Y, Xia W, Zhao C, et al. Recent advances in bacterial therapeutics based on sense and response. Acta Pharm Sin B 2023;13:1014–27. https://doi.org/10.1016/j.apsb.2022.09.015.

[130] Clairmont C, Lee KC, Pike J, Ittensohn M, Low KB, Pawelek J, et al. Biodistribution and genetic stability of the novel antitumor agent VNP20009, a genetically modified strain of Salmonella typhimurium. J Infect Dis 2000;181:1996–2002. https://doi.org/10.1086/315497.

[131] Toso JF, Gill VJ, Hwu P, Marincola FM, Restifo NP, Schwartzentruber DJ, et al. Phase I study of the intravenous administration of attenuated Salmonella typhimurium to patients with metastatic melanoma. J Clin Oncol 2002;20:142–52.

[132] Chirullo B, Ammendola S, Leonardi L, Falcini R, Petrucci P, Pistoia C, et al. Attenuated mutant strain of Salmonella Typhimurium lacking the ZnuABC transporter contrasts tumor growth promoting anti-cancer immune response. Oncotarget 2015;6:17648–60.

[133] Kim J-E, Phan TX, Nguyen VH, Dinh-Vu H-V, Zheng JH, Yun M, et al. Salmonella typhimurium suppresses tumor growth via the pro-inflammatory cytokine interleukin-1β. Theranostics 2015;5:1328–42. https://doi.org/10.7150/thno.11432.

[134] Canale FP, Basso C, Antonini G, Perotti M, Li N, Sokolovska A, et al. Metabolic modulation of tumours with engineered bacteria for immunotherapy. Nature 2021;598:662–66. https://doi.org/10.1038/s41586-021-04003-2.

[135] Duong MT-Q, Qin Y, You S-H, Min -J-J. Bacteria-cancer interactions: bacteria-based cancer therapy. Exp Mol Med 2019;51:1–15. https://doi.org/10.1038/s12276-019-0297-0.

[136] Raman V, Van Dessel N, Hall CL, Wetherby VE, Whitney SA, Kolewe EL, et al. Intracellular delivery of protein drugs with an autonomously lysing bacterial system reduces tumor growth and metastases. Nat Commun 2021;12:6116. https://doi.org/10.1038/s41467-021-26367-9.

[137] Zheng JH, Nguyen VH, Jiang S-N, Park S-H, Tan W, Hong SH, et al. Two-step enhanced cancer immunotherapy with engineered Salmonella typhimurium secreting heterologous flagellin. Sci Transl Med 2017;9:eaak9537. https://doi.org/10.1126/scitranslmed.aak9537.

[138] Jiang S-N, Park S-H, Lee HJ, Zheng JH, Kim H-S, Bom H-S, et al. Engineering of bacteria for the visualization of targeted delivery of a cytolytic anticancer agent. Mol Ther 2013;21:1985–95. https://doi.org/10.1038/mt.2013.183.

[139] Nguyen D-H, Chong A, Hong Y, Min -J-J. Bioengineering of bacteria for cancer immunotherapy. Nat Commun 2023;14:3553. https://doi.org/10.1038/s41467-023-39224-8.

[140] Ho JML, Miller CA, Parks SE, Mattia JR, Bennett MR. A suppressor tRNA-mediated feedforward loop eliminates leaky gene expression in bacteria. Nucleic Acids Res 2021;49:e25. https://doi.org/10.1093/nar/gkaa1179.

[141] Greco FV, Pandi A, Erb TJ, Grierson CS, Gorochowski TE. Harnessing the central dogma for stringent multi-level control of gene expression. Nat Commun 2021;12:1738. https://doi.org/10.1038/s41467-021-21995-7.

[142] Fernandez-Rodriguez J, Voigt CA. Post-translational control of genetic circuits using Potyvirus proteases. Nucleic Acids Res 2016;44:6493–502. https://doi.org/10.1093/nar/gkw537.

[143] Ganai S, Arenas RB, Forbes NS. Tumour-targeted delivery of TRAIL using Salmonella typhimurium enhances breast cancer survival in mice. Br J Cancer 2009;101:1683–91. https://doi.org/10.1038/sj.bjc.6605403.

[144] Li L, Pan H, Pang G, Lang H, Shen Y, Sun T, et al. Precise thermal regulation of engineered bacteria secretion for breast cancer treatment in vivo. ACS Synth Biol 2022;11:1167–77. https://doi.org/10.1021/acssynbio.1c00452.

[145] Wang L, Cao Z, Zhang M, Lin S, Liu J. Spatiotemporally controllable distribution of combination therapeutics in solid tumors by dually modified bacteria. Adv Mater 2022;34:2106669. https://doi.org/10.1002/adma.202106669.

[146] Abedi MH, Yao MS, Mittelstein DR, Bar-Zion A, Swift MB, Lee-Gosselin A, et al. Ultrasound-controllable engineered bacteria for cancer immunotherapy. Nat Commun 2022;13:1585. https://doi.org/10.1038/s41467-022-29065-2.

[147] Anderson NM, Simon MC. Tumor microenvironment. Curr Biol 2020;30:R921–5. https://doi.org/10.1016/j.cub.2020.06.081.

[148] Qin W, Xu W, Wang L, Ren D, Cheng Y, Song W, et al. Bacteria-elicited specific thrombosis utilizing acid-induced cytolysin A expression to enable potent tumor therapy. Adv Sci 2022;9:2105086. https://doi.org/10.1002/advs.202105086.

[149] Chien T, Harimoto T, Kepecs B, Gray K, Coker C, Hou N, et al. Enhancing the tropism of bacteria via genetically programmed biosensors. Nat Biomed Eng 2022;6:94–104. https://doi.org/10.1038/s41551-021-00772-3.

[150] Panteli JT, Forbes NS. Engineered bacteria detect spatial profiles in glucose concentration within solid tumor cell masses. Biotechnol Bioeng 2016;113:2474–84. https://doi.org/10.1002/bit.26006.

[151] Chowdhury S, Castro S, Coker C, Hinchliffe TE, Arpaia N, Danino T. Programmable bacteria induce durable tumor regression and systemic antitumor immunity. Nat Med 2019;25:1057–63. https://doi.org/10.1038/s41591-019-0498-z.

[152] Gurbatri CR, Lia I, Vincent R, Coker C, Castro S, Treuting PM, et al. Engineered probiotics for local tumor delivery of checkpoint blockade nanobodies. Sci Transl Med 2020;12:eaax0876. https://doi.org/10.1126/scitranslmed.aax0876.

[153] Savanur MA, Weinstein-Marom H, Gross G. Implementing logic gates for safer immunotherapy of cancer. Front Immunol 2021;12:780399. https://doi.org/10.3389/fimmu.2021.780399.

[154] Anderson JC, Voigt CA, Arkin AP. Environmental signal integration by a modular AND gate. Mol Syst Biol 2007;3:133. https://doi.org/10.1038/msb4100173.

[155] Chabloz A, Schaefer JV, Kozieradzki I, Cronin SJF, Strebinger D, Macaluso F, et al. Salmonella-based platform for efficient delivery of functional binding proteins to the cytosol. Commun Biol 2020;3:1–11. https://doi.org/10.1038/s42003-020-1072-4.

[156] Park S-H, Zheng JH, Nguyen VH, Jiang S-N, Kim D-Y, Szardenings M, et al. RGD peptide cell-surface display enhances the targeting and therapeutic efficacy of attenuated *Salmonella*-mediated cancer therapy. Theranostics 2016;6:1672–82. https://doi.org/10.7150/thno.16135.

[157] Critchley-Thorne RJ, Stagg AJ, Vassaux G. Recombinant Escherichia coli expressing invasin targets the Peyer's patches: the basis for a bacterial formulation for oral vaccination. Mol Ther 2006;14:183–91. https://doi.org/10.1016/j.ymthe.2006.01.011.

[158] Bai F, Li Z, Umezawa A, Terada N, Jin S. Bacterial type III secretion system as a protein delivery tool for a broad range of biomedical applications. Biotechnol Adv 2018;36:482–93. https://doi.org/10.1016/j.biotechadv.2018.01.016.

[159] Reeves AZ, Spears WE, Du J, Tan KY, Wagers AJ, Lesser CF. Engineering Escherichia coli into a protein delivery system for mammalian cells. ACS Synth Biol 2015;4:644–54. https://doi.org/10.1021/acssynbio.5b00002.

[160] Lynch JP, Goers L, Lesser CF. Emerging strategies for engineering Escherichia coli Nissle 1917-based therapeutics. Trends Pharmacol Sci 2022;43:772–86. https://doi.org/10.1016/j.tips.2022.02.002.

[161] Hyun J, Jun S, Lim H, Cho H, You S-H, Ha S-J, et al. Engineered attenuated Salmonella typhimurium expressing neoantigen has anticancer effects. ACS Synth Biol 2021;10:2478–87. https://doi.org/10.1021/acssynbio.1c00097.

[162] Lynch JP, González-Prieto C, Reeves AZ, Bae S, Powale U, Godbole NP, et al. Engineered Escherichia coli for the in situ secretion of therapeutic nanobodies in the gut. Cell Host Microbe 2023;31:634–49, e8. https://doi.org/10.1016/j.chom.2023.03.007.

[163] Danino T, Prindle A, Kwong GA, Skalak M, Li H, Allen K, et al. Programmable probiotics for detection of cancer in urine. Sci Transl Med 2015;7:289ra84. https://doi.org/10.1126/scitranslmed.aaa3519.

[164] Cooper RM, Wright JA, Ng JQ, Goyne JM, Suzuki N, Lee YK, et al. Engineered bacteria detect tumor DNA. Science 2023;381:682–86. https://doi.org/10.1126/science.adf3974.

[165] Elliott KT, Neidle EL. Acinetobacter baylyi ADP1: transforming the choice of model organism. IUBMB Life 2011;63:1075–80. https://doi.org/10.1002/iub.530.

[166] Leventhal DS, Sokolovska A, Li N, Plescia C, Kolodziej SA, Gallant CW, et al. Immunotherapy with engineered bacteria by targeting the STING pathway for anti-tumor immunity. Nat Commun 2020;11:2739. https://doi.org/10.1038/s41467-020-16602-0.

[167] Luke JJ, Piha-Paul SA, Medina T, Verschraegen CF, Varterasian M, Brennan AM, et al. Phase I study of SYNB1891, an engineered E. coli Nissle strain expressing STING agonist, with and without atezolizumab in advanced malignancies. Clin Cancer Res 2023;29:2435–44. https://doi.org/10.1158/1078-0432.CCR-23-0118.

[168] Synlogic. A Phase 1, Open-label, Multicenter Study of SYNB1891 Administered by Intratumoral Injection to Patients With Advanced/Metastatic Solid Tumors and Lymphoma Alone and in Combination With Atezolizumab. clinicaltrials.gov 2021.

[169] Russell BJ, Brown SD, Siguenza N, Mai I, Saran AR, Lingaraju A, et al. Intestinal transgene delivery with native *E. coli* chassis allows persistent physiological changes. Cell 2022;185:3263–77, e15. https://doi.org/10.1016/j.cell.2022.06.050.

[170] Puurunen MK, Vockley J, Searle SL, Sacharow SJ, Phillips JA, Denney WS, et al. Safety and pharmacodynamics of an engineered E. coli Nissle for the treatment of phenylketonuria: a first-in-human phase 1/2a study. Nat Metab 2021;3:1125–32. https://doi.org/10.1038/s42255-021-00430-7.

[171] Staedtke V, Roberts NJ, Bai R-Y, Zhou S. Clostridium novyi-NT in cancer therapy. Genes & Diseases 2016;3:144–52. https://doi.org/10.1016/j.gendis.2016.01.003.

[172] Diaz LA, Cheong I, Foss CA, Zhang X, Peters BA, Agrawal N, et al. Pharmacologic and toxicologic evaluation of C. novyi-NT spores. Toxicol Sci 2005;88:562–75. https://doi.org/10.1093/toxsci/kfi316.

[173] Sidney Kimmel Comprehensive Cancer Center at Johns Hopkins. Phase I Safety Study of Clostridium Novyi-NT Spores in Patients With Treatment-Refractory Solid Tumor Malignancies. clinicaltrials. gov 2016.

[174] Gurbatri CR, Radford GA, Vrbanac L, Im J, Thomas EM, Coker C, et al. Engineering tumor-colonizing E. coli Nissle 1917 for detection and treatment of colorectal neoplasia. Nat Commun 2024;15:646. https://doi.org/10.1038/s41467-024-44776-4.

[175] Zhang H, Dai Z, Wu W, Wang Z, Zhang N, Zhang L, et al. Regulatory mechanisms of immune checkpoints PD-L1 and CTLA-4 in cancer. J Exp Clin Cancer Res 2021;40:184. https://doi.org/10.1186/s13046-021-01987-7.

[176] Ciardiello D, Vitiello PP, Cardone C, Martini G, Troiani T, Martinelli E, et al. Immunotherapy of colorectal cancer: challenges for therapeutic efficacy. Cancer Treat Rev 2019;76:22–32. https://doi.org/10.1016/j.ctrv.2019.04.003.

Nazlıcan Tunç, Mehmet Emin Bakar, Burak Çalışkan,
Tolga Tarkan Ölmez, and Urartu Özgür Şafak Şeker

Chapter 5
New generation cellular engineering for living therapeutics

Abstract: This chapter explores the progress and possibilities of cutting-edge cellular engineering for therapeutic purposes. This statement underscores the fundamental change in medical treatments, moving away from conventional medicines toward living therapeutic systems. It emphasizes the significance of synthetic biology in developing self-replicating systems that have the ability to independently diagnose, treat, and cure diseases. The study focuses on several important topics, including the development of genetic circuits for medical purposes, the use of the human microbiome as a therapeutic platform and its impact on health and disease, the promise of microbiome engineering, and novel approaches for delivering drugs and treating diseases. The chapter also explores the use of biomaterials such as biofilms and biomineralization in treatments, as well as the importance of cross-disciplinary collaboration in the development of therapeutic approaches. The incorporation of these advanced technology offers potential solutions for unfulfilled medical requirements and revolutionizes treatment strategies for a range of illnesses.

5.1 Introduction

5.1.1 A paradigm shift

Over the last century, there have been significant discoveries and developments in medicine, but there still need to be unmet clinical needs across numerous human diseases [1]. With the development of technology, medical trends have begun to shift away from traditional therapeutic methods to living therapeutic systems. This shift began with natural proteins, used as the first biological therapeutics. These are involved in polyclonal antibodies and purified porcine insulin. Then, it evolved into monoclonal antibodies. Another evolution is living therapeutics [2]. Compared to traditional methods, living therapeutics are independently self-replicating factories that

Nazlıcan Tunç, Mehmet Emin Bakar, Urartu Özgür Şafak Şeker, UNAM–Institute of Materials Science and Nanotechnology, Bilkent University, Ankara 06800, Turkey
Burak Çalışkan, Department of Biomedical Engineering, TOBB University of Economics and Technology, Ankara 06510, Turkey
Tolga Tarkan Ölmez, Department of Biomedical Engineering, Başkent University, Ankara 06790, Turkey

https://doi.org/10.1515/9783111329499-005

produce and pump out therapeutics within the body and use a cell scaffold to diagnose, treat, or cure a disease [3]. In addition, it has several advantages, such as decreased production cost, reduced side effects, and lowered required dose [4].

5.1.2 Cellular engineering in therapeutics

Synthetic biology allows us to build genetic circuits we can design, characterize, test, and analyze and integrate into our living system [5, 6]. A genetic circuit comprises three elements: an input module that detects signals and converts them into molecular signals, an operation module that processes the signal and decides on the cell's response, and an output module that translates the processed signal into the desired cellular reaction, often through genes encoding biological effectors. In therapeutics, output modules are enzymes, cytokines, or cell receptors [1].

Selecting a chassis is crucial; the most used are bacteria, eukaryotic cells, and bacteriophages [7]. Among them, *Escherichia coli* (*E. coli*) is the most commonly used chassis due to its safety as an organism to work with, convenient genetic manipulation capabilities, well-studied background, and abundance in the human gut [8]. These features make it a dependable option for delivering drugs and metabolites to specific body tissues, such as tumor cells [9]. *E. coli* can be engineered to secrete biologically active therapeutic molecules, like human interleukin-10, which can downregulate proinflammatory cytokines in conditions such as inflammatory bowel disease (IBD) [10]. Additionally, *E. coli* can function as a carrier system for the in situ delivery of therapeutic molecules in the gut, presenting innovative approaches for treating various diseases [11]. Therefore, it is considered a good option for living therapeutics.

5.2 The human microbiome as a therapeutic platform

A deeper understanding of the impact of the human microbiome on health has propelled the field in the last few decades [12]. However, the unambiguous importance of the microbiome had already been speculated shortly after the isolation of *E. coli* in 1885 [13, 14]. Although investigations on what would later be termed microbiome, which, contrary to the widely held belief not coined by Joshua Lederberg [15], had started in the nineteenth century, scientific insights on human microbiome have gained unprecedented acceleration only lately to the point that the microbiome was even defined as the "last organ" [16]. Still, the definition of the microbiome or the difference between microbiome and microbiota often needs to be clarified. The human microbiome is defined as all members of bacteria, fungi, archaea, as well as algae, protists, viruses, and phages in and on the body, plus their collective genomes, biological processes, and interactions both between each other and their human host [17]. The coevolution of the microbiome with its human host has affected many systems in the human body, including but not limited to metabolism, immunity, diseases, disorders, and even behavior. The health of an individual seems to be related to the microbiome of the gut, particularly in such a way that a disruption of the equilibrium leads to dysbiosis [18], a broken link between the gut microbiome and host, which would unavoidably affect the healthy state of the organism. The so-called nonequilibrium state of the microbiome–human host relation could arise from a disorder in the human host, eventually leading to a disorder in the microbiome or vice versa. Understanding the importance of the microbiome in health and disease has led to breakthrough discoveries. Although microbiome research is much older than many would realize, the field is witnessing an ardent expansion as synthetic biology tools are being applied to reprogram the microbiome, which could potentially lead to transformative therapeutic platforms for a remarkably large range of disorders from cancer to autoimmunity to antimicrobial resistance (AMR). The therapeutic potential of the human microbiome is vast because there are few places, if any, in the gastrointestinal (GI), respiratory, and urogenital tracts that are untouched by microorganisms.

The process of intervening in the microbiome to rebalance its healthy interaction with the human body is known as "microbiome engineering" [19]. There are many different approaches to engineering the human microbiome as a therapeutic modality. Nevertheless, harnessing the potential of microbiomes for therapeutic purposes could be broadly categorized into two parts. The first part involves conventional methods such as altering the composition of the microbiome through the introduction of probiotics or specific beneficial strain(s), supplementation of different prebiotics, or transferring strains between individuals, a remarkably successful example of which is fecal microbiota transplantation (FMT) for recurrent *Clostridium difficile* infection [20, 21]. The second part takes advantage of cutting-edge synthetic biology approaches that are being used, though hitherto only as preclinical research, to engineer a specific strain of the microbiome for a more precise and targeted intervention. On the one hand, the conventional method of using well-known harmless strains to reshape the microbial balance has been known and applied for decades. On the other hand, in recent years, state-of-the-art techniques armed with synthetic biology have been used to program microorganisms for much more challenging problems, such as addressing hard-to-treat diseases through living bacterial therapeutics [22].

5.2.1 Microbiome of the human body as potential living therapeutics

The human body's microbiome holds great potential to be harnessed as a therapeutic platform because of the collective effect of many microorganisms on health. Although it might sound surprising, until recently, the number of cells in a human body was contentious without consensus among the scientific community. A proper examination has settled the discussion lately. It is now assumed that there are about 36, 28, and 17 trillion cells in an adult male, an adult female, and a child, respectively [23]. There has been an extended standing comparison between the number of bacterial and human cells, and a widely believed misconception refers to the ratio of bacterial to human cells as 10:1. It is now accepted that the actual ratio is probably close to 1:1 [24]. Setting the ratio right would be needed because modulating the microbiome toward an expected therapeutic outcome may depend on the extent of the bacterial community. Different organs host different strains and compositions of microbial communities, which are implicated in diseases, as will be summarized below before moving to microbiome engineering applications.

The gut microbiome is the most studied one in the human body. It has been known for some time that restoring the gut microbiome is one way of harnessing the microbiome as a therapeutic platform. The reason for balancing the microbiome as a therapeutic intervention is that the microbiome occupies a niche in the gut, which helps inhibit the colonization of pathogenic bacteria. A perturbation to the microbiome shifts the permeability of the gut microenvironment, which enables pathogens to invade parts of the gut that they are not even supposed to reach in the first place [25, 26]. As an example of dysbiosis, IBD is associated with gut microbiome imbalance. IBD comprises a set of long-lasting and relapsing diseases. The two most common examples of IBD are ulcerative colitis (UC) and Crohn's disease (CD), which are thought to arise from changes in gut microbiome specificity [27]. It is assumed that the reduction of some species in the gut is related to IBD progression; for instance, a decrease in the abundance of *Faecalibacterium prausnitzii* is frequently observed in IBD cases [28, 29]. Targeting *F. prausnitzii* could thus be a potential therapeutic intervention for IBD treatment (Figure 5.1). Like IBD, irritable bowel syndrome (IBS) [30], celiac disease [31], and colorectal cancer [32] are also known to be in connection with the gut microbiome. For instance, the ratio of *Firmicutes* (Bacillota) to *Bacteroidetes* (Bacteroidota) is increased in IBS patients [33, 34]. A loss of abundance of beneficial bacteria affects epithelial barrier function, which may be restored through an intervention to increase the abundance of Bacteroidota.

Although not well-characterized compared to the gut microbiome, the nasal microbiome is thought to affect several diseases in the respiratory system. As clinical evidence, dysbiosis of the sinonasal microbiome is implicated in patients suffering from chronic rhinosinusitis [35]. Another affliction related to the respiratory system microbiome is chronic obstructive pulmonary disease (COPD). It was found that the lung microbiome of COPD patients is distinguished from that of healthy individuals, with differences in the abundance of *Prevotella*, *Streptococcus*, and *Moraxella* species [36]. Similarly, the development of asthma is known to be associated not only with lung microbiome alone but also with the communication of the gut–lung axis [37]. The asthma example is one of the deeper aspects of the interactions between micro-

Healthy state · Crohn's disease

No inflammation ─── ─── *F.prausnitzii* ─── Inflammation throughout the colon

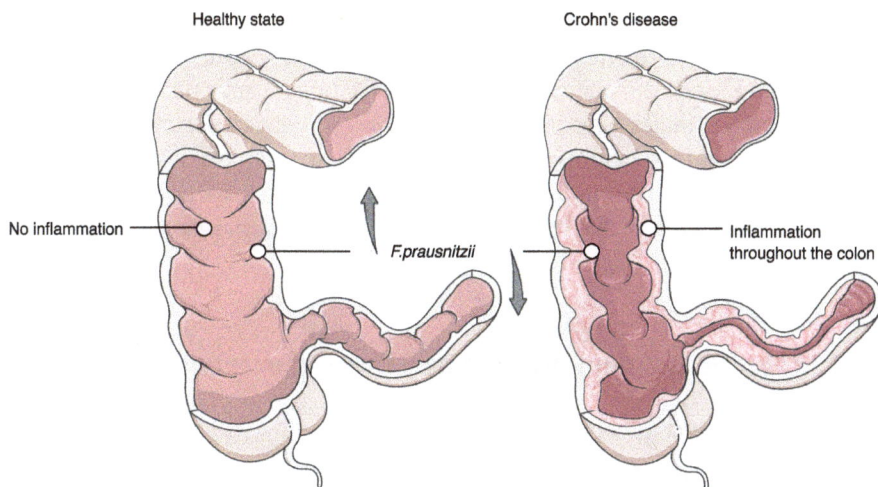

Figure 5.1: The abundance of microbial community differs between healthy and disease states. A decrease in abundance of *F. prausnitzii* is observed in the colon during inflammatory bowel disease. Conversely, an increase in *F. prausnitzii* is a signature of remission (schematics are prepared using images from Servier Medical Art under CC BY 4.0 license).

biomes of different systems in the body, a prominent one being the gut–brain axis [38].

It is unsurprising for the skin to host a diverse array of microorganisms because of its exposure directly to the environment. The somewhat surprising point is the stability of the skin microbiome [39]. The interference to the skin microbiome could be a viable option for several cases. An example is a eukaryotic polyomavirus known to cause Merkel cell carcinoma, a form of skin cancer [40]. Another example is *Propionibacterium acnes*, an anaerobic bacterium believed to be the causative factor of the skin condition of acne vulgaris [41]. Probably a more relevant and commonly observed example of skin microbiome-related pathogenesis is atopic dermatitis (AD), commonly known as "eczema." It is observed in AD cases that the microbiome's diversity is decreased, and the abundance of *Staphylococcus aureus* increases accordingly [42, 43]. Although AD is a chronic disease without a definitive cure, it might be possible to target *S. aureus* as a therapeutic option to address AD severity, for which promising results using *R. mucosa* were reported [44].

The urobiome, the microbiome of the urinary tract, is the least known microbiome compared to other body parts. Although the urobiome is understudied and underexplored, it is no less important than the other microbial communities in the human body. The bladder was once considered sterile. However, it is now known that it hosts microorganisms that contribute to both healthy and disease states [45–47]. Urobiome plays a vital role in some diseases, specifically recurrent urinary tract infections, which, as its name indicates, becomes resistant to antibiotics and is known through clinical data to be related to urobiome dysbiosis [48]. Although the first-line treatment option is antibiotics, a post-antibiotic era is approaching, and addressing UTIs needs to include the microbiome as a biological therapeutic platform. A drawback in urobiome studies is the problem of obtaining and identifying microorganisms from standard urinary cultures; however, enhanced quantitative urine culture improves the procedure [49]. In addition, sequencing is an indispensable method for bacterial identification, and it includes 16S rRNA, shotgun metagenomic, and whole genome sequencing [50–52].

5.2.2 Probiotic supplementation as a living therapeutic platform

The close relationship between microorganisms and the human body implies an untapped potential to harness specific strains for therapeutic interventions against various diseases. As mentioned above, many parts of the human body have host-specific bacterial communities whose imbalances lead to various disorders. Interfering the abundance of harmful strains in favor of beneficial ones is one way of using the microbiomes as a therapeutic platform. It is worth noting that every individual hosts her/his own unique microbiome for which many factors, such as geography and genetics, are in play. The rebalancing of the ratio between different microbial communities could be achieved through probiotic supplementation. Harnessing living therapeutics has a longer history than many would think.

A well-known example is the isolation of what was later named *E. coli* Nissle 1917 (EcN) during the First World War from a soldier who had been to the Balkans, where Shigella was rampant. The immunity of the solider against Shigella led Alfred Nissle to suspect the existence of an antagonistic strain in the sample. A few years back, Alfred Nissle came up with the concept of antagonistic activity after observing the inhibition of some coliform pathogens and cocultured *Salmonella* strains in the presence of specific *E. coli* isolates [53]. EcN is still used as a probiotic supplementation (e.g., Mutaflor®) to address dysbiosis-related conditions, including IBDs [54].

Probiotic supplementation often improves gut microbiota, immune system responses, and female reproductive system in healthy adults; however, the benefits are generally short-lived [55]. Because the main focus of probiotic supplementation is the patients with specific disorders, there have been randomized controlled trials to investigate the effect of microbiome change in obesity [56, 57] and in UC or CD [58]. A randomized controlled trial from 2023 shows that in one group of patients with metabolic syndromes such as obesity, high blood pressure, and alike, the triglycerides and diastolic blood pressure significantly improved; however, the other group in the study experienced the opposite, meaning that their metabolic syndrome markers worsen after probiotic supplementation, which is thought to be related with the diet of the individuals [59]. Similar studies imply that despite being a widely adopted therapeutic intervention, the use of probiotics warrants caution. Probiotic supplementation may not be beneficial unless prudence is exercised. One reason is that taking too much of a good thing could be harmful when, for instance, many strains are outcrowded by a limited number of beneficial strains. Similarly, specific allergies could be triggered, or antibiotic resistance genes could be affected, particularly in at-risk populations with severe disease and compromised immune systems if they rely on probiotic supplementation [60]. Another important aspect is the use of antibiotics along with probiotics. In three groups of healthy adults who took antibiotics for a week, it was observed that the first group of individuals who also took probiotics for 1 month after antibiotics did much worse in terms of regaining a healthy microbiome compared to both the control group and the other group who underwent autologous fecal microbiome transplantation [61].

Harnessing probiotics as a living therapeutic platform could be coupled to prebiotics, which are nondigestible substrates to foster microbial community growth. One step forward, postbiotics, which are beneficial products that microorganisms synthesize, could also be coupled with probiotic treatments for a more robust option. For instance, microbiota-targeted dietary intervention through plant-based fibers or fermented foods improves healthy adults' immune systems. Fermented foods are shown to enhance the microbiome's diversity and decrease markers related to inflammation [62]. Moreover, prebiotics could enhance the effect of immune checkpoint inhibitors used as immunotherapy for certain types of cancer. It is known that gut microbiome composition affects the outcome of checkpoint blocker programmed cell death protein-1 (PD-1) [63–65]. It could be helpful to modulate the gut microbiome with probiotics, prebiotics, and FMT to increase the efficacy of immune checkpoint therapy [66, 67]. In conclusion, using probiotics as living therapeutics along with prebiotics and postbiotics could lead to an enhanced therapeutic outcome, probably also by-passing probiotic-only related risks at the same time.

While increasing the abundance of specific strains for a therapeutic outcome is viable, doing the opposite to decrease and eliminate some strains is sometimes needed. Although antibiotics and vaccines are used to eliminate or prevent the colonization of harmful bacteria, AMR is a pressing challenge that inexorably arises sooner or later after an antibiotic is introduced [68–71]. A potentially robust countermeasure against bacteria is to harness phages that invade specific strains of bacteria and either lyse the host or integrate their genes into the genome to be activated later [72]. Contrary to most antibiotics, which indiscriminately kill bacteria, phages are highly precise and allow the survival of beneficial bacteria in the body [73, 74]. It is also possible to engineer phages used as therapeutics to eliminate specific bacteria through double-strand breaks in the host genome by programmed CRISPR-Cas9 (clustered regularly interspaced short palindromic repeats) [75]. It was shown that delivery of CRISPR-Cas9 through phages or bacterial plasmids enables targeted knockdown of bacteria discriminately [76]. In addition, another option to interfere with pathogenic bacteria is to use bacterial secretion systems, which are syringe-like protein complexes used to inject proteins into a target cell [77–79]. Armed with the CRISPR-Cas system, it was demonstrated that microbial communities or targeted strains in the gut or only a specific pathogenic strain could be engineered on demand [80–82].

Engineering or supplementing a single strain could confer short-term benefits in specific cases; however, because the effect of microorganisms on the human body arises from their act as a community, it is imperative to modulate a variety of microbial consortia to observe long-term benefits. Transferring or supplementing a cocktail of defined or undefined microbial communities is more effective than probiotics comprising a limited number of specific strains. As mentioned above, a well-known application of microbial transfer en masse between individuals is FMT. Although the most promising feature of FMT is to treat recurrent *C. difficile* cases [83], for which the US Food and Drug Administration (FDA) approved Rebyota in 2022 as the first fecal microbiota product, there are challenges in using FMT as a therapeutic platform [84]. This includes finding a donor with suitable microbial consortia and eliminating the risk of transferring harmful or antibiotic-resistant bacteria. In addition, the members of the virome of the gut are mostly phages, most of which are little known [85]. Notably, such was the case during the COVID-19 pandemic; there is also the risk of transferring pathogenic viruses and microbial communities [86]. For these reasons, using natural or synthetic microbial cocktails containing spore-forming Firmicutes is a viable option that bypasses the risks associated with FMT. A randomized, double-blind, placebo-controlled phase III clinical trial found that a natural cocktail of bacterial spores could address recurrent *C. difficile* infection [87]. In contrast to natural microbial consortia, it might be more effective to rationally prepare a synthetic microbial consortia, a cocktail of microorganisms containing predetermined members that would allow distinguishing the effect of different strains on therapeutic outcomes [88, 89]. Animal models showed that rationally designed bacterial consortia could address IBD and restore intestinal homeostasis [90].

5.2.3 Cellular programming with synthetic biology for living therapeutics

Harnessing wild-type beneficial bacteria from the human microbiomes has been shown to provide at least short-term treatment options, as mentioned in the previous sections; however, to achieve long-term success, engineering microorganisms equipped with sense-and-respond gene circuits to external stimuli is needed [4, 91–93]. Apart from FMT, there is currently a lack of successful translation of microbial therapies from bench to bedside [94]. There are excellent proof-of-concept studies, but the clinical outcomes are often discouraging. Many promising engineered therapies have been discontinued, such as engineered EcN to treat hyperammonemia (ClinicalTrials.gov ID:NCT03447730). The chassis organism was an orally delivered engineered bacterium designed for hepatic insufficiency and cirrhosis patients to convert ammonia to L-arginine. To work only in the low-oxygen environment of the gut, the metabolic pathway was controlled through an anaerobic-inducible fnrS promoter. In addition, an auxotrophy system based on thymidine dependency was built to ensure biosafety, thereby avoiding the survival of genetically modified organisms outside the intended environment [95]. Although a phase I clinical trial did show efficacy, the following phase II trial did not provide benefit in cirrhosis patients compared to placebo. Similarly, EcN was engineered to address phenylketonuria (PKU), an in-born metabolic syndrome. PKU is a metabolic disorder characterized by the inactivity or lack of phenylalanine hydroxylase, an enzyme degrading phenylalanine (Phe) that, if it builds up in the blood, causes severe neurological damage. EcN was designed with three gene circuits: a Phe transporter gene under the control of an anaerobic-inducible fnrS promoter (PfnrS), a phenylalanine ammonia-lyase (PAL) gene (stlA), and L-amino acid deaminase (LAAD) gene (pma) under arabinose-inducible araC promoter (ParaC). The engineered bacteria were able to convert Phe into trans-cinnamate, which in turn metabolized to hippurate (HA), a metabolite that is excreted in the urine, thereby providing a convenient biomarker [96]. Following promising results, an enhanced version of the engineered cell was also designed, which led to a two-fold increase in in vivo PAL activity [97]. Despite successful phase I and II results, it was concluded that the primary endpoint could not be met based on the results of a phase III trial, which led to the discontinuation of the therapeutic strain (ClinicalTrials.gov ID:NCT04984525).

CRISPR-based approaches are also being widely used for microbiome engineering [98]. Because Cas9 induces double-strand breaks in DNA, directly harnessing Cas9 is not viable to engineer microorganisms because complex DNA repair mechanisms are unavailable in prokaryotes, in contrast to engineering human cells where homology-directed repair could be harnessed [99]. Instead, CRISPR interference (CRISPRi) using deactivated Cas9 could be deployed to specifically knock down target genes for microbiome engineering [70]. For instance, a CRISPRi library was constructed to interfere with any chosen gene in *Bacillus subtilis*, which could be used to test a variety of different effects in the GI tract through engineered bacteria [100]. In contrast to interfering gene expression, CRISPR-mediated transcriptional activation (CRISPRa) could also be used. Direct application of the CRISPR system into target microorganisms in situ is challenging because, contrary to in vitro experiments, transferring a plasmid or Cas proteins in bacteria would be a significant hurdle. The gut microbiome, in particular, is not limited to its genomes or plasmids; it exchanges genetic material through horizontal gene transfer (HGT) [101]. Depending on the function of transferred genes through HGT, this could be a double-edged sword. It is also possible to use metagenomic engineering to hijack bacterial injection systems and transfer target genes into either gram-positive or gram-negative bacteria in the mammalian gut. However, the transferred material becomes undetectable in days [102].

Synthetic biology approaches are being deployed to address the toughest challenges in many fields. One of the challenges that synthetic biology could come to rescue is antibiotic resistance. Antibiotic-induced dysbiosis affects microbiome alterations in the human body and paves the way for multidrug-resistant pathogens. To address the problem of antibiotic-related depletion of beneficial bacteria, *Lactococcus lactis* was engineered to metabolize b-lactams through secretion and extracellular assembly of heterodimeric b-lactamase to avoid disruptions of commensal bacteria in the gut. In a mouse model, oral supplementation of the engineered bacteria was shown to protect the gut microbiome without affecting serum ampicillin level [92]. Probiotic strains could also be modified as living biosensors to detect pathogen-derived metabolites and respond to inform as a point-of-need readout. *L. lactis* was engineered to specifically detect quorum-sensing molecules synthesized by *Vibrio cholerae* in the gut and express enzymatic reporter that could be colorimetrically probed from fecal samples. The engineered strain was found to suppress cholera in mice through lactic acid production [103]. An attractive option to harness microorganisms as living diagnostics within the gut environment is to engineer a memory system to report disease-related changes such as inflammation. An example of such a system was constructed as a two-part module with a trigger element. The repressor Cro protein from bacteriophage l was transcribed from an anhydrotetracycline-inducible promoter (Ptet), and the memory element was derived from the cI/Cro region. In the presence of anhydrotetracycline (aTC), the Cro state is stably turned on; conversely, in the absence of aTC, the cI state is on. The memory system remembers the state it experienced even after the withdrawal of the trigger element, in this case, aTC. Because the Cro-ON state could be interrogated by a b-galactosidase reporter linked to Cro, it is possible to extract information about the inflammation state if it is experienced [104]. This concept could be extended to detect specific small molecules for probing disease-related signals produced in the gut microenvironment. Another inflammation-sensing bacteria for the gut was designed to sense and respond to metabolite tetrathionate, which is synthesized during inflammation. Similar to the previous examples of aTC induction, Cro protein was used. However, its expression was controlled through the *ttR/S* gene from *S. typhimurium* using the ttRBCA promoter (PttRBCA). Once tetrathionate induces Cro expression, activation of a b-galactosidase is detected through fecal samples even though tetrathionate no longer exists in the environment. The stability of the system was enhanced when integrated into the chromosome of *E. coli*, which was shown to work in vivo for 6 months in a mouse model [105].

Since the end of the nineteenth century, it has been known that bacterial infections could, in some cases, clear tumor cells in cancer patients if they are infected with certain strains. In the early days of microbiology, live bacteria were recognized as anticancer agents known as "Coley toxins" following the observation of William Coley that one of his patient's cervical sarcoma disappeared after a streptococcal infection [106]. From those early days, it was realized that bacteria could infiltrate tumors and might be used as living therapeutics [107]. It is now known that there is a human tumor microbiome (Figure 5.2). This means that not only specific strains could enter into tumors, but different intracellular bacteria can also infiltrate cancer cells in a tumor-type-specific manner [108]. Having realized that wild-type intracellular bacteria have a natural proclivity for immune privilege in tumor cores that immune cells fail to access, it was hoped that further engineering of selected strains could be armed with genetic circuits to act as a Tro-

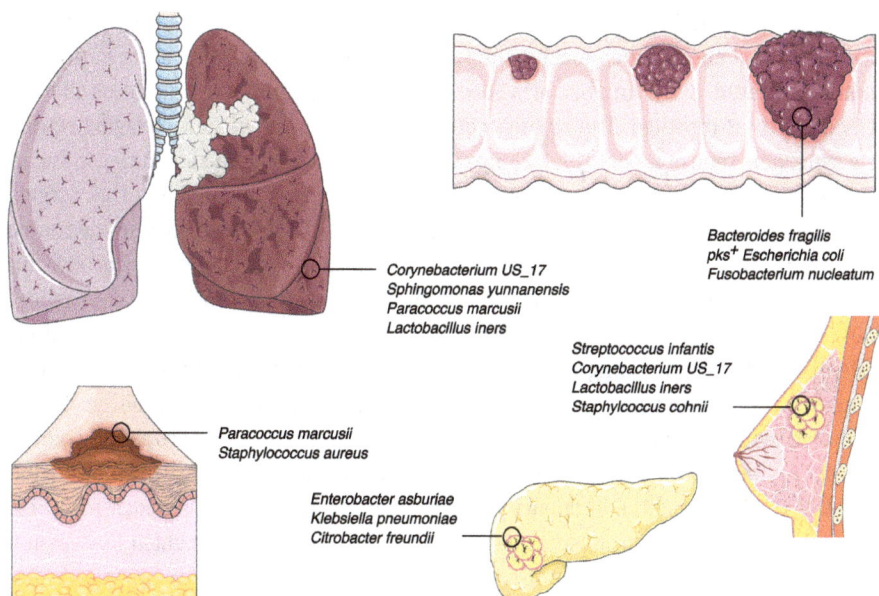

Figure 5.2: Human tumor microbiome for lung, colon, breast, pancreas, and skin cancers. Each organ hosts a different tumor microbiome. It is known that there are intratumoral pathogenic bacteria that survive in tumor cells because of tumor-provided immune privilege; however, some members of the commensal bacteria also convert into a pathogenic state and infiltrate into tumors. (Clockwise from top right) Colon cancer-related bacteria include *B. fragilis*, pks+ *E. coli*, *F. nucleatum*; breast cancer is known to host a variety of bacteria involving *S. infantis*, *F. nucleatum*, *P. argentinensis*, *E. cloacae*, *Acinetobacter* US_424, *Corynebacterium* US_1715, *L. iners*, and *S. cohnii*; 30–50% of pancreatic tumors is known to host *E. asburiae*, *K. pneumoniae*, *C. freundii*; melanoma, a type of skin cancer, was found to contain two strains *P. marcusii* and *S. aureus*; lung cancer microbiome includes *Corynebacterium*, *S. yunnanensis*, *P. marcusii*, *L. iners* as well as *N. macacae*, *R. mucosa*, *K. pneumoniae*, and *C. freundii* (schematics are prepared using images from Servier Medical Art under CC BY 4.0 license).

jan horse and discriminately kill cancer cells from inside while healthy cells are untouched. Although it sounds attractive, there are only a limited number of strains that are well-known enough to be predictable.

Moreover, most intratumoral bacteria must be more studied, hitherto cultivable, or genetically intractable. However, it is possible to engineer workhorse microorganisms to enhance their tumor tropism. Once inside the tumor or within the tumor microenvironment, these engineered cells could compute cellular signals to produce an output accordingly that could determine the fate of tumors. Alternatively, bacteria can be reprogrammed to harness quorum sensing for synchronous lysing at a threshold population density and deliver genetically encoded cargo into the tumor microenvironment.

A genetic circuit with positive and negative feedback comprising luxI promoter that controls its activator acyl-homoserine lactone (AHL) as a quorum sensing molecule and a bacterial lysis gene (φX174E) was constructed. Once the AHL concentration reaches a threshold, lysis protein E is expressed, and an encoded cargo of a proapoptotic peptide, a chemokine, and hemolysin, which acts as a pore-forming antitumor toxin, is released. Surviving bacteria seed the growing population, and an oscillatory delivery cycle is repeated. The circuit-engineered bacteria showed encouraging results in a syngeneic mouse model [109]. Nonpathogenic *E. coli* was also shown to conduct a similar cancer-targeting mechanism using quorum sensing. A nanobody that targets antiphagocytic receptor CD47 was encoded in bacteria to be delivered within the tumor microenvironment upon lysis. It was shown that these tumor-colonizing bacteria increased T-cell infiltration into tumors and tumor regression in a mouse tumor model [110]. Along a similar line, probiotic EcN was engineered to lyse and intratumoral release of nanobodies targeting programmed cell death-ligand 1 (PD-L1) and cytotoxic T lymphocyte-associated protein-4 (CTLA-4) for immune checkpoint inhibition. When combined with probiotically synthesized cytokine granulocyte-macrophage colony-stimulating factor (GM-CSF), enhanced therapeutic activity was observed in a murine model [111]. Increasing the efficiency of immune checkpoint inhibitors such as PD-L1-blocking antibodies is a promising concept used with microbiome-enabled therapeutic intervention. In this respect, tumor-colonizing EcN was engineered to convert ammonia, a metabolic waste of tumors, into L-arginine. Increased L-arginine concentration in tumors enhances the activity of tumor-infiltrating T cells, which helps clear the tumors in synergy with checkpoint inhibition [112].

Engineered bacteria could also detect cancer rather than target it directly. An example is the CATCH platform (cellular assay for targeted, CRISPR-discriminated HGT), which uses bacteria to detect tumor DNA. *Acinetobacter baylyi* was used to collect tumor DNA as it is known for its tendency to integrate cell-free DNA into its genome, which is called "natural competence." *A. baylyi* was designed to gain antibiotic resistance after integrating a specific sequence from the mutated KRAS gene. Integrating the system in a mouse tumor model verified the successful signaling of cancer through the detection of *A. baylyi* after antibiotic treatment [113].

One common drawback of living therapeutics is their inability to colonize target tissue, such as the GI tract. Although it is preferable that engineered bacteria be cleared from the gut after a limited period for safety reasons, long-term activity of the engineered organisms is needed for enhanced therapeutic effect. Despite the advantage of EcN in colonizing the gut around two weeks, the modification of the bacterial surface with thiol groups for covalent attachment to disulfide-rich mucus is a viable option to perform longer activity [114]. The downside of this strategy is the loss of surface modification with each generation, for which further engineering might confer an inherent ability of surface modification in a chosen organism. Wild-type probiotic bacteria are often inept when complex intervention is needed.

Moreover, there could be cases that require intravenous bacterial injection. In such cases, engineered bacteria are detected through specialized immune system cells. Therefore, bacteria need to be camouflaged from immune cells, which were demonstrated using a surface-expressed inducible coating for EcN. The system allows immune evasion until the shedding of the capsule in nontarget tissues, where immune cells clear the bacteria rapidly [115]. Rather than using live bacterial cells, bacteria-derived particles have also been demonstrated as potential therapeutics. Outer membrane vesicles (OMV) are nanoparticles produced by Gram-negative bacteria such as *Akkermansia muciniphila*, a member of the human gut microbiome and is known to regulate intestinal homeostasis [116]. Engineered *E. coli*-derived OMVs were shown to induce interferon-g-mediated durable antitumor response without notable side effects in a murine model [117]. In a step forward, in situ production of tumor-antigen-bearing arabinose-inducible *E. coli* OMVs in the gut stimulated dendritic cells and provided immunization against rechallenge in a mouse model [118].

An intriguing option is to modify human microbiome-related bacteria to synthesize artificial nanoparticles that could target and deliver payloads into specific tissues. One such nano-sized system is a minicell, which is an achromosomal cell that arises from mutations in the minB locus in *E. coli* upon aberrant cell division [119]. These minicells could be modified as delivery modules because of their intrinsic ability to absorb cytoplasmic proteins and plasmids. It was shown that gene circuits could be integrated into minicells to detect small molecules that induce green fluorescent protein (GFP) expression [120], deliver chemotherapeutic drugs [121], or convert salicylate into anticancer molecule catechol guided with nanobodies for cancer biomarkers [122]. The increasing complexity of living therapeutics has been recently leveraged by combining chimeric antigen receptor (CAR)-T cells with engineered bacteria. CAR-T cells are approved therapeutic modified T cells that provide robust treatments against hematologic malignancies. However, solid tumors have not yielded to CAR-T cells for various reasons, such as on-target off-tumor toxicity and poor infiltration of T cells into tumors. The ProCAR platform (Figure 5.3) was recently shown to target solid tumors effectively through engineered probiotic-guided T cells. Attenuated EcN was designed to express ProTag, which comprises a heparin-binding domain along with GFP. The former tags the extracellular matrix (ECM) in the tumor microenvironment, and the latter is engineered to be recognized by CAR-T cells. Engagement with GFP is one of the system's key features because GFP plays a crucial role as a bio-orthogonal marker. Furthermore, an enhanced version of the platform, ProCombo, adds an improved engineered probiotic for releasing chemokine 16 (CXCL16K42A), which triggers both inflammation and infiltration of CAR-T cells into tumors. The combined results imply reduced tumor volume, enhanced survival, increased T-cell infiltration, and tumor clearance in humanized and immunocompetent mouse models of leukemia, colorectal, and breast cancer [123]. It is important to note that there is significant potential for improvement of the platform to target immunologically cold tumors that are elusive to the immune system.

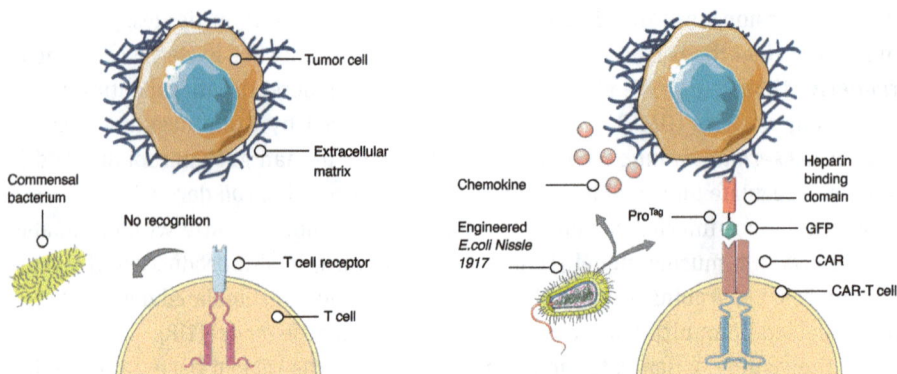

Figure 5.3: Probiotic-guided chimeric antigen receptor T-cell (ProCAR) platform. (Left) If T cells cannot recognize tumors, aberrant division of cells goes unchecked. Although some pathogenic bacteria have been known to counteract tumor growth, ability of commensal bacteria to interfere is little to none. (Right) In ProCAR platform, engineered *E. coli* Nissle 1917 releases GFP-heparin binding domain fusion to target ECM. Tagging ECM in this way allows CAR-T-cell migration to tumors because the CAR is specific for GFP recognition. Bio-orthogonal nature of GPF avoids the risk of off-target activity. An enhanced version of ProCAR adds a second genetic circuit to release chemokine 16 (CXCL16K42A) that triggers additional immune responses to help an elevated attack on tumor cells (schematics are prepared using images from Servier Medical Art under CC BY 4.0 license).

5.3 Engineered living materials (ELMs) as therapeutics

Engineered living materials (ELMs) represent a fusion of genetic engineering and materials science, designed to create adaptive, responsive, and multifunctional materials. These materials integrate living cells, such as bacteria or mammalian cells, with synthetic elements, enabling them to grow, self-repair, and interact dynamically with their environment. The potential applications of ELMs span a broad spectrum, including therapeutic delivery, biosensing, tissue engineering, and environmental monitoring (Figure 5.4). By utilizing the unique properties of biofilms, such as curli and TasA biofilms from *E. coli* and *B. subtilis*, researchers can engineer systems for targeted drug delivery, pathogen detection, and self-healing materials. Advanced applications include the development of wearable biosensors, living tattoos, and ingestible devices for real-time health monitoring. Additionally, the process of biomineralization, inspired by natural mineral formation, is harnessed to create materials for bone regeneration and cancer therapy. Through the integration of living cells with synthetic frameworks, ELMs offer innovative solutions that could transform medical treatments, environmental remediation, and bioelectronic interfaces, highlighting the ver-

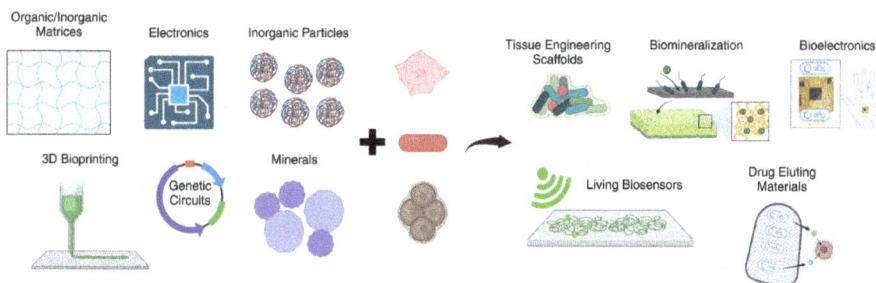

Figure 5.4: Diverse components and applications of engineered living materials (ELMs) in biotechnology and materials science. It showcases the integration of organic/inorganic matrices, electronic components, and inorganic particles to enhance ELM properties. Techniques like 3D bioprinting and genetic circuits enable precise fabrication and programmable behaviors. Applications depicted include tissue engineering scaffolds, living biosensors, and drug-eluting materials for controlled therapeutic delivery, biomineralization, and bioelectronics, emphasizing the potential of ELMs in creating smart, adaptive, and multifunctional materials for various therapeutic uses.

satility and promise of this emerging field. The unique aspects and novel application potentials of ELMs are discovered in this chapter by cutting-edge examples.

5.3.1 Engineering biofilm systems as therapeutic ELMs

5.3.1.1 Curli biofilms of *E. coli*

The curli operon is a genetic module found in many Enterobacteriaceae, including *E. coli*, which encodes the components necessary to produce curli fibers. Curli are extracellular amyloid fibers that play crucial roles in biofilm formation and host interaction. The curli operon consists of several genes, including *csgA* and *csgB*, which encode the major and minor curli subunits, respectively, and *csgD*, a transcriptional regulator that activates curli production. Additional components like csgE, csgF, and csgG are involved in the secretion and assembly of curli fibers on the cell surface. In biomedical fields, curli fibers have garnered interest due to their potential applications in biomaterial engineering, as they can serve as scaffolds for tissue engineering, platforms for drug delivery, and tools for vaccine development. The unique properties of curli, such as their ability to form stable and biofunctional amyloid structures, make them promising candidates for various biomedical applications [124, 125].

CsgA, the main building block of curli, can be genetically fused to various protein and peptide units to create functionalized biopolymers, such as creating environmentally responsive biofilm glues using genetically engineered *E. coli*. These living glues incorporate the CsgA-Mfp3s fusion protein, which combines the self-assembling amyloid nanofibers of curli proteins with the strong adhesive properties of mussel foot proteins, resulting in enhanced adhesion. The engineered biofilms can perform various mechanical tasks on demand, such as capturing microspheres from solutions to form composite coatings and autonomously repairing damage in microfluidic devices. The glues are triggered by different stimuli, including chemicals, light, and blood. Notably, the blood-responsive glue system can detect blood leakage and initiate repair autonomously, making it particularly promising for medical applications. Examples like this highlight the potential of combining these living glues to revolutionize material science by providing smart, self-repairing materials for medical use [126].

Researchers explore the application of synthetic biology to develop biofilms that utilize the *csgA* gene to produce therapeutic agents for targeted and sustained drug delivery, such as GI disorders. In a prominent study, the therapeutic potential of engineered *E. coli* Nissle 1917 (EcN) biofilms was shown by producing curli fibers fused with trefoil factors (TFFs) to treat IBD. By engineering the curli operon, specifically utilizing the *csgA* gene to produce TFFs, known to promote epithelial barrier function and repair, the researchers created biofilms that adhere to intestinal ulcers and provide localized therapy. In vivo, experiments using a mouse model demonstrated that these biofilms when orally administered, could effectively colonize the gut, reduce inflammation, and promote mucosal healing. The engineered bacteria showed enhanced intestinal residence time and produced significant therapeutic effects, including reduced disease activity and improved histological outcomes. This approach highlights the potential of synthetic biology in developing novel treatments for GI disorders by leveraging the inherent properties of biofilms for targeted and sustained drug delivery [127].

Indirect disease management can also be conducted by water treatment using genetically engineered biofilms, thereby controlling the spread of viral infections. *E. coli* can be modified to produce biofilms that include CsgA proteins fused with an influenza-virus-binding peptide, referred to as CsgA-C5. The genetic modification involves fusing the CsgA protein with a peptide known to bind the hemagglutinin protein of the influenza virus, thereby creating the CsgA-C5 fusion protein. The C5 peptide was previously identified by phage display and shown to have a high affinity for hemagglutinin, a key glycoprotein on the surface of influenza viruses. When CsgA is fused with the C5 peptide, the resultant protein can self-assemble into stable amyloid fibers while presenting the virus-binding sites externally. This configuration ensures that the fibers can effectively interact with and capture influenza virus particles from water [128].

The engineering of biofilm systems for a versatile and efficient method to achieve site-specific, covalent attachment of proteins and enzymes, facilitating stable and reusable platforms for various biotechnological applications, is also an active area of research. One such study focuses on developing catalytic biofilms using *E. coli* expressing CsgA-SpyTag. In this system, curli nanofibers, which are naturally produced by *E. coli*, are engineered to include SpyTag sequences. These sequences allow for site-specific covalent attachment of enzymes that are fused to SpyCatcher, a protein that binds to SpyTag with high affinity. This innovative method facilitates the stable immobilization of enzymes onto the curli nanofibers, creating biofilms with catalytic properties [129]. Building on this foundational work, the application of curli fibers for enzyme immobilization was further explored. In their study, they used the CsgA-SpyTag system to attach enzymes involved in the conversion of starch to trehalose [130]. The immobilized enzymes retained their activity and stability, demonstrating the effectiveness of curli nanofibers as a platform for biocatalysis. This approach not only enhances the efficiency of enzymatic reactions by providing a stable and reusable enzyme platform but also opens up possibilities for designing complex multienzyme systems for various biomedical applications.

Bacterial biofilms can also be engineered for bioelectronics applications. In this manner, researchers examined the effects of integrating aromatic amino acids, known for their electronic conductivity, into the amyloid-like fibers formed by the CsgA. By measuring the conductivity properties of different aromatic amino acids, tyrosine and tryptophan were identified as most effective in promoting electronic conductivity due to their ability to form delocalized π clouds similar to those found in conductive polymers. To achieve this, *E. coli* was genetically programmed to produce curli fibers embedded with the conductive peptide motifs. This modification involved adding sequences of three aromatic amino acids to the fiber-forming peptides, enhancing their stacking behavior and, consequently, their conductivity [131]. The genetically engineered *E. coli* then synthesized biofilms with significantly improved conductive properties. These conductive biofilms have the potential to interface with electrodes and connect bacterial populations to electronic devices, opening up new possibilities in bioelectronics and biosensing applications.

The curli operon, particularly the *csgA* gene, has shown significant potential in biomedical applications. Engineered *E. coli* producing curli fibers fused with functional proteins have been developed for various uses. These include biofilm glues for autonomous repair, targeted drug delivery systems for GI disorders, and water treatment solutions to capture influenza viruses. These innovations highlight the versatility and promise of curli-based technologies in advancing medical and biotechnological fields.

5.3.1.2 TasA biofilms of *B. subtilis*

B. subtilis biofilms are composed of protein and exopolysaccharide (EPS) structures. The main building block of the protein components is the TasA protein. Recently, TasA has garnered significant attention in the field of biofilm engineering due to its unique properties. It plays a crucial role in biofilm formation and stability. The molecular architecture of TasA filaments and their importance in biofilm scaffold formation and stability is well-studied [132]. The self-assembling nature of TasA fibers contributes to the structural integrity of biofilms, making them robust and resistant to environmental stresses [133]. Furthermore, TasA has been identified as an amyloid-like protein, forming filaments with dye-binding activity (such as Congo Red) characteristic of amyloid proteins [133, 134].

Engineering the TasA biofilm of *B. subtilis* can be beneficial for various biomedical applications, including drug delivery systems and tissue engineering. By utilizing the genetic tool of TasA for the display of heterologous proteins, researchers have demonstrated the potential of TasA biofilms in presenting antigenic peptides for therapeutic purposes. First, they demonstrated the widespread deployment of the recombinant tasA subunits within *B. subtilis* biofilms by a fusion protein, TasA-mCherry. They further enhanced the expression of tasA by depleting the *sinR* gene, which is the repressor for both tasA and eps structures. Subsequently, fusion proteins combining TasA with antigenic peptides from the *Echinococcus granulosus* parasite, such as paramyosin and tropomyosin, were successfully expressed within the biofilms. The study also demonstrated that recombinant endospores of *B. subtilis* maintained their biophysical and morphological properties, making them viable delivery agents for heterologous proteins. This capability was confirmed by showing that the spores could withstand harsh conditions and maintain their integrity [135]. Hence, tasA can function as a delivery tool for therapeutic agents.

There is also active research for the development of flexible platforms for creating living materials using genetically engineered *B. subtilis* embedded into the materials along with their biofilm structures. TasA amyloid machinery can be engineered to create fusion proteins that self-assemble into functional extracellular nanofibers. These biofilms exhibit tunable physicochemical properties and can be precisely fabricated into various 3D structures using 3D printing and microencapsulation techniques. Key applications demonstrated include fluorescence detection, enzymatic bioremediation, and the assembly of inorganic nanoparticles, highlighting the biofilms' multifunctional and environmentally responsive capabilities. Technical challenges, such as the secretion of large proteins and controlled processing of complex materials, can be solved by leveraging the TasA fusion proteins' high tolerance to functionalization. The engineered biofilms exhibit viscoelastic behaviors akin to hydrogels, making them suitable for diverse applications in biomaterials, biotechnology, and biomedicine [136]. The 3D-printed and functionalized *B. subtilis* biofilms display the potential of these biofilms in creating self-regenerating, long-lasting, and environmentally adaptive materials, providing a new approach to the development of dynamic and multifunctional living materials with important implications for nanomanufacturing.

5.3.1.3 *L. lactis* biofilms

L. lactis, a gram-positive bacterium known for its probiotic properties, has gained attention in biofilm engineering for biomedical applications, particularly in the context of stem cell research. The ability of *L. lactis* to form biofilms on various surfaces, such as glass and synthetic polymers, has paved way for its utilization as a long-term substrate for mammalian cell culture, including human mesenchymal stem cells (hMSCs) [137, 138].

Early studies employed lactococcal promoters and secretion signals to express a fibronectin fragment (FNIII 7–10) on the *L. lactis* bacterial membrane, without biofilm formation [139]. The FNIII 7–10 was chosen because it incorporates both the PHSRN synergy sequence and the RGD binding motifs, which are crucial for promoting cell adhesion and signaling. These motifs interact with integrins, a family of transmembrane receptors that mediate cell attachment to the ECM and initiate various signaling pathways. By displaying this specific fibronectin fragment on the membrane of *L. lactis*, the bacteria can effectively mimic the natural ECM, enhancing cell adhesion, spreading, and signaling processes such as focal adhesion formation and phosphorylation of focal adhesion kinase. This targeted approach ensures that the bacterial biointerface provides the necessary cues to direct mammalian cell behavior in a controlled and dynamic manner, which is essential for applications in regenerative medicine [139]. Another type of cell integrated into the same system for the direction of differentiation was myoblasts. When seeded on the *L. lactis* biofilms, C2C12 myoblasts showed improved adhesion, spreading, and formation of focal adhesion plaques. The biofilms maintained stability and viability over several days, ensuring continuous exposure of the FN fragment to the myoblasts. Therefore, the biofilms of modified *L. lactis* can provide a stable and dynamic interface, promoting higher levels of myogenic differentiation than traditional substrates like fibronectin and collagen-coated surfaces [137]. FNIII7–10 fragment of human fibronectin was also used for creating a living biointerface capable of supporting and directing hMSC differentiation into osteoblasts. The engineered *L. lactis* biofilms were stable and viable for up to 28 days, facilitating long-term hMSC culture and differentiation when combined with bone morphogenetic protein 2 (BMP-2). The hMSCs showed enhanced adhesion, spreading, and osteogenic differentiation, demonstrated by increased osteocalcin expression and phosphate deposition, on the biofilms compared to control surfaces [138]. Therefore, the engineered *L. lactis* biofilms can mimic the dynamic nature of the ECM, providing biochemical cues to direct cell behavior, capable of presenting various biochemical signals to control cell fate in a spatiotemporal manner.

For wound-healing applications, hydrogel-encapsulated engineered microbial consortia, including *L. lactis*, have been investigated for their potential to promote skin wound healing [140]. Specifically, CXCL12-secreting engineered *L. lactis* has been utilized to construct living materials that have shown efficacy in enhancing wound healing in rat-skin defect models. The hydrogel-encapsulated engineered microbial consortium (HeEMC) designed as a photoautotrophic living material includes genetically modified *Synechococcus elongatus* PCC7942, which produces sucrose through photosynthesis, and *L. lactis*, which uses the sucrose to grow and secrete functional biomolecules such as CXCL12. The hydrogel matrix, composed of PEGDA and chitosan, provides a supportive environment that enhances cell proliferation and ensures the stability and functionality of the microorganisms. The HeEMC demonstrated significant efficacy in promoting wound healing in a rat model, reducing the wound area significantly. The system addresses the challenge of maintaining microbial activity in the dry and nutrient-poor environment of the epidermis, making it a promising approach for treating skin wounds and potentially other epidermal conditions [140].

5.3.1.4 Bacterial cellulose

Bacterial cellulose (BC), a biopolymer with many beneficial properties and wide-ranging applications, is primarily synthesized by strains of *Gluconacetobacter xylinus* and *Komagataeibacter rhaeticus*. These bacteria are distinguished by their capability to produce cellulose with unique physicochemical characteristics, differentiating it from plant cellulose [141]. *G. xylinus*, previously known as *A. xylinum*, has been thoroughly investigated for its cellulose-synthesizing mechanisms. The cellulose generated by these bacteria, commonly known as "bacterial cellulose" (BC), is noted for its high crystallinity, tensile strength, water absorption capacity, biocompatibility, resistance to degradation, and low solubility, making it highly beneficial for various engineered tissue applications [142].

K. rhaeticus, another species within the *Gluconacetobacter genus*, has been identified as a model organism associated with plants for BC biosynthesis. This bacterium, formerly known as *Gluconacetobacter xylinus* ATCC 53582, has been investigated for its cellulose production in response to the phytohormone ethylene, which not only enhances cellulose production but also regulates gene expression within the cellulose synthesis operon [143]. Through a deep understanding of the factors influencing cellulose production, genetic modifications, and innovative fabrication techniques, these bacteria continue to drive advancements in the utilization of BC for diverse biomedical applications.

Developing methods for producing BC using 3D printing technology, alongside other functional applications is an active area of research. A specialized biocompatible ink composed of composed of hyaluronic acid (HA), K-carrageenan, and fumed silica, termed "Flink," was used to embed *A. xylinum* during the 3D printing. This technique aimed for precise control over the spatial distribution and concentration of the bacteria within the printed structures, ensuring their survival and activity. The printed BC demonstrated high mechanical strength and the ability to form complex shapes, making it suitable for biomedical applications such as tissue engineering [144]. BC production by 3D printing has been advanced through the integration of different bioinks such as by embedding cellulose-producing bacteria into silicone-based granular gels. This approach circumvents traditional limitations in shaping BC into complex geometries, as the viscoelastic properties and gas permeability of the silicone gel support bacterial growth and cellulose formation post-printing. *G. xylinus* was used to study the mechanical properties and growth behavior of the resulting cellulose structures. The granular gel facilitates the maintenance of living bacterial cells within the printed structures, allowing for self-regeneration and self-healing of the cellulose networks. This method provides a platform for creating sustainable materials with potential applications in biomedicine due to its ability to produce ordered, self-repairing cellulose structures [145].

The use of BC as a regenerative tool was also explored with BC spheroids as modular building blocks for ELMs, produced through shaking cultures of *K. rhaeticus*. These spheroids can be genetically engineered to incorporate functional properties, making them suitable for constructing 3D shapes and patterned materials. The spheroids' ability to self-assemble and regenerate damaged areas reflects their potential to create self-healing materials [146].

A recent trend in the field of ELMs is the production of the materials by cocultures. A coculture system of genetically engineered *Saccharomyces cerevisiae* yeast and *K. rhaeticus* bacteria benefits from the bulk production of BC by *K. rhaeticus* and secretion of specific enzymes that enhance the functionality of the BC by the yeast [147]. This coculture system, termed "Syn-SCOBY" (synthetic symbiotic culture of bacteria and yeast), mimics the natural division of labor seen in microbial communities, where different organisms contribute to different functions within the material. These ELMs can be functionalized with biosensing, where the yeast can detect and respond to environmental stimuli by producing fluorescent proteins, making them candidates for pollutant detection and environmental monitoring. In biocatalysis, the materials can autonomously degrade pollutants or synthesize chemicals, exemplified by yeast secreting β-lactamase to break down antibiotics in wastewater. The self-healing properties of these materials enable them to regenerate after damage, maintaining their functionality over time. Additionally, their high biocompatibility and customizability make them suitable for biomedical applications, such as wound dressings and tissue engineering scaffolds. Overall, the coculture approach combines synthetic biology and materials science to develop sustainable, adaptive, and programmable materials with a wide range of therapeutic uses [148].

5.3.2 Chemical and biomarker detection with ELMs

ELMs represent an emerging field that combines genetic engineering with organic and inorganic materials to develop responsive materials capable of adapting to various stimuli. These materials can be integrated living cells with synthetic components, offering platforms for creating stimuli-responsive devices [149]. By utilizing biological material assembly processes within designed materials, ELMs exhibit growth, repair, and adaptive behaviors in response to external cues, making them versatile for a wide range of applications [147] such as the detection of biomarkers for diagnostics and on-demand release of therapeutics within responsive ELM scaffolds.

Stretchable patches for chemical and biomarker detection require bacteria inside an organic or inorganic material. Traditional methods often fail to preserve cells over long periods, leading to leakage or loss of functionality. In a study aiming to solve this problem, genetically engineered bacterial cells were integrated into stretchable and robust hydrogel–elastomer hybrids [150]. Their approach aimed to harness the capabilities of living cells while maintaining their viability and functionality as wearable devices. The hydrogel component of the hybrid provides a sustainable supply of water and nutrients to the encapsulated cells, while the elastomer component ensures sufficient air permeability to maintain cell viability. These elastomeric hybrid materials frequently undergo deformations since they are designed to be worn on the skin. Bending may cause the leakage of bacteria from the material, making it nonfunctional, but the elastomeric nature of the ELM also circumvents the leakage problem. This capability is demonstrated in various formats, including skin patches and gloves, highlighting the versatility and potential of these living sensors in practical applications. Furthermore, the creation of interactive genetic circuits within freestanding devices was implemented. These circuits involve bacterial strains that communicate by the diffusion of signaling molecules through the hydrogel. For example, a transmitter strain produces a quorum-sensing molecule when induced, which then triggers a receiver strain to produce GFP. This setup allows for the study of cellular signaling cascades and demonstrates the potential for more complex interactions within living devices [150].

A prominent benefit of integrating engineered bacteria is the ability to design multiplexed detection systems for many chemicals or biomarkers with genetic circuits. Multiplexed detection can be achieved by using the same ELM with different engineered bacteria strains producing the same response to different signals [150, 151]. As an example, bacterial sensors can be programmed to respond to specific chemical inducers, such as IPTG, AHL, and rhamnose, by producing GFP as an individual response to each, allowing for the simultaneous detection of multiple chemicals. On the other hand, multiplexed detection systems may also rely on spatial logic gates. In such systems, bacteria are modified with specific sensing circuits, and different circuits are patterned with techniques such as 3D printing to form Boolean logic gates, which output visible fluorescent light. The use of 3D printing enables the creation of complex, responsive constructs and allows the printing of large-scale (up to 3 cm), high-resolution (30 µm) structures that maintain cell viability and functionality. The hydrogel matrices provide a biocompatible environment, sustaining the cells with nutrients and allowing for the diffusion of signaling molecules. These ELMs attached to the skin, also named "living tattoos," expand the toolbox by creating customized, responsive living materials and devices [151].

Pathogen sensing and elimination can be combined on the same biosensing ELM as well. Platform for adhesion-mediated trapping of cells in hydrogels (PATCH) is a good example of such systems [152]. This platform integrates genetically engineered *E. coli* that express a glucose-binding adhesion protein derived from an *Antarctic bacterium*. The adhesion stably anchors the bacteria within dextran-based hydrogels that have large pore sizes (10–100 µm), which significantly reduces the leakage of the bacteria into the environment by up to 100-fold compared to conventional designs. The hydrogel matrix supports the diffusion of macromolecules, thereby preserving the repertoire of molecules available for cellular communication and biosensing functions. One of the notable applications demonstrated with PATCH is the creation of living materials that can sense and respond to specific environmental stimuli by secreting therapeutic proteins. For instance, the engineered *E. coli* within the hydrogel can produce and secrete lysostaphin, an antimicrobial enzyme that specifically targets and kills *S. aureus*, including methicillin-resistant *S. aureus* [152]. This capability is particularly valuable in medical and environmental contexts where the rapid and specific detection and neutralization of pathogens are critical.

Besides skin patches for biosensing and pathogen elimination, a cutting-edge approach is to develop in situ devices for monitoring. This is important to circumvent traditional invasive techniques. In such systems, bacterial biosensors can be combined with electronic devices for signal detection and transmission. The ingestible micro-bioelectronic device (IMBED) integrates genetically engineered probiotic bacteria with advanced miniaturized electronics to create a system capable of in situ biomolecular detection for GI health. The engineered bacteria are designed to be heme-sensitive, responding to the presence of blood by producing luminescence, which is detected by the device's electronics. The readings can be wirelessly transmitted the data to an external receiver, such as a smartphone or a computer. This capability was successfully demonstrated in a swine model, where the IMBED device accurately diagnosed GI bleeding. The system's design ensures that the bacteria remain viable and functional within the harsh environment of the GI tract. The researchers demonstrated the device's flexibility by integrating alternative biosensors capable of detecting other clinically relevant biomarkers, such as thiosulfate, which is elevated in conditions like gut inflammation, and AHL, a molecule associated with bacterial communication. This modularity allows the IMBED device to be adapted for various diagnostic applications, potentially expanding its use to a wide range of GI conditions [153].

Merging genetic engineering with organic and inorganic materials to create responsive, adaptable systems is a promising approach in biosensing applications. These materials offer potential for a wide range of applications, including diagnostics, therapeutic delivery, biosensing, and pathogen elimination. By integrating living cells into robust and flexible materials, researchers have developed innovative solutions like stretchable patches for chemical detection, self-repairing biofilms, and ingestible devices for in situ biomolecular monitoring. These advancements highlight the versatility and promise of ELMs in medical and environmental contexts, paving the way for future innovations in responsive, smart materials that can interact dynamically with their environment to address complex challenges in healthcare and beyond.

5.3.3 Biomineralization as a tool for therapeutic ELMs

Biomineralization is a complex process observed in nature, where living organisms influence the precipitation of mineral materials. This phenomenon is crucial for the formation of biominerals, which possess unique properties and structures due to organic–inorganic interactions orchestrated by living organisms [154]. In nature, biomineralization is essential for the creation of various materials, such as bone apatite, where organic molecules from the ECM guide the structuring of the mineral component [155]. The taxonomic distribution of biominerals often involves phosphate and carbonate salts of calcium, combined with organic polymers like collagen and chitin to provide structural support to bones and shells [156]. By mimicking natural biomineralization processes, researchers aim to develop synthetic inorganic materials with specific structures and properties, inspired by the precision and efficiency of biologically controlled mineralization [157]. The biomineralized ELMs may offer unique therapeutic advantages.

The development of novel living material platforms that utilize synthetic biology for the biomineralization of silica and hydroxyapatite were researched, aiming for potential therapeutic applications in dental and bone regenerative medicine [158]. An example platform uses *E. coli* bacteria engineered to display a peptide (R5-FP) known for its ability to precipitate silica on its surface by its antigen system. This peptide tag is used by diatoms to synthesize their biomineralized cell walls. Displaying it on the bacterial surface allows for the rapid and economical production of silica nanoparticles without the need for laborious protein purification processes. One key application demonstrated is in endodontics, where the silica produced by the bacteria enhances the osteogenic differentiation of human dental pulp stem cells (hDPSCs), promoting earlier calcium crystal deposition [158]. Similarly, the biomineralization of calcium phosphate crystals through protein–protein interactions, particularly osteocalcin (OCN) and osteopontin (OPN), was also shown [159]. These proteins can be utilized to tailor the nucleation and growth of hydroxyapatite (HAP) crystals, which are essential for bone regeneration and repair. By fine-tuning the concentrations and interactions of OCN and OPN, the researchers achieved controlled HAP crystal formation, leading to improved design of biomimetic matrices that mimic mammalian hard tissue environments, which support osteoblast differentiation to osteocytes [159]. These advancements hold the potential for enhancing bone healing and regeneration.

A precise spatial control over HA formation is needed for bone regeneration applications. In this manner, living composite materials using light-inducible bacterial biofilms engineered to promote hydroxyapatite (HA) mineralization were developed [160]. Researchers used *E. coli* biofilms functionalized with curli fibers (CsgA) fused with mussel foot protein (Mfp) analogs, specifically Mfp3S-pep, to induce and control HA mineralization. The genetic design allowed for precise spatial control of biofilm formation and subsequent mineralization through light exposure. The engineered bacteria, termed "light receiver-CsgA–Mfp3S-pep," were able to produce amyloid fibers under blue light, which served as templates for HA deposition. The ability to create living materials that mimic the gradient mechanical properties of natural bone tissues can facilitate the regeneration of complex tissue interfaces, such as those between bone and cartilage. Moreover, the living composites can retain their viability and responsiveness to environmental cues, which is crucial for dynamic and adaptive tissue repair processes [160].

Biomineralization can also be used for specific targeting of cancer via drug delivery using biomineralized bacteria coated with metal-organic frameworks. In a recent study, *E. coli* was engineered with a zeolitic imidazolate framework-8 (ZIF-8) layer, coloaded with the photosensitizer chlorin e6 (Ce6) and the chemotherapeutic drug doxorubicin (DOX), through a one-step in situ method. This *E. coli*@ZIF-8/C&D hybrid maintains bacterial viability and tumor selectivity, leveraging the natural tumor-homing capabilities of *E. coli*. Upon exposure to near-infrared laser irradiation, the *E. coli*@ZIF-8/C&D generated reactive oxygen species, enhancing the cytotoxic effects against cancer cells through a synergistic chemo-photodynamic therapy approach. This combination therapy showed significant inhibition of tumor growth in mouse models, with minimal side effects [161].

The potential of bacterial systems as noninvasive imaging tools is also an emerging concept. As an example, genetically engineered bacteria can be functionalized to accumulate magnetite nanocrystals for use as magnetic resonance imaging (MRI) contrast agents, with promising therapeutic applications in cancer diagnostics and treatment. Researchers developed two systems: extracellular magnetite accumulating bacteria (ECMAB) and intracellular magnetite accumulating bacteria (ICMAB). In the ECMAB system, *E. coli* bacteria were engineered to display the Mms6 protein on their surface, which facilitated the extracellular formation of magnetite nanoparticles. This approach improved MRI contrast performance due to the enhanced magnetic properties of the synthesized nanomagnets. For the ICMAB system, the bacteria were modified to express a combination of Mms6, ferroxidase, and a metal-binding peptide, allowing them to synthesize magnetite nanoparticles intracellularly. These intracellular nanomagnets exhibited strong MRI contrast capabilities. The engineered bacteria not only improved imaging but also offered the potential for hyperthermia treatment by targeting tumor sites with localized heating [162].

In summary, biomineralization in nature is a process where living organisms orchestrate the formation of mineral materials through organic–inorganic interactions. This phenomenon is essential for the creation of biominerals with unique properties and structures, showcasing the remarkable adaptability and efficiency of biological systems in mineral formation. By studying and mimicking natural biomineralization processes, researchers aim to develop advanced materials and technologies inspired by the precision and sophistication of biologically controlled mineralization.

5.4 Cross-disciplinary partnership in therapeutic development

Interdisciplinarity is described as the ability to combine and integrate expertise across different scientific disciplines [163]. This expertise may include scientific approaches, methodologies, protocols, and insights from an expert's research area(s). An interdisciplinary approach may be required to tackle a complex problem that has proven unsolvable by the concepts and knowledge of a single scientific branch. Although the scientific knowledge from many disciplines can be amalgamated into a single structure containing mixed elements from each discipline, discrete insights and approaches from each discipline can only enrich the toolbox of capabilities that a group of researchers can have. The complex research problem that one seeks to address usually requires innovative thinking. Innovative thinking is often a product of multiple minds focusing on a problem to build novel concepts on top of each other. This synergistic approach can help develop a new way of thinking about that complex research problem [164–166]. A good example of innovative thinking might be the invention of microarray technology. DNA array chip fabrication, nucleic acid hybridization, and analysis tool development require combined expertise in engineering, semiconductor technologies, chemical biology, biochemistry, molecular genetics, and bioinformatics [167]. Fabrication includes the design and development of masks that allow deprotection of probes by UV exposure, whereas hybridization is an optimization problem that can be solved by efficient and high-purity synthesis of oligo probes [168]. Developing new tools and computer program packages to analyze microarray data involves both handling big data and determining optimal parameters for quality control of microarray data [169]. Innovative thinking is often a product of two or more disciplines having very low alignment working toward a shared goal. Disciplines having few similarities are inherently risky in interdisciplinary research, while they have a better chance to result in a product that has innovative and emergent properties.

Living therapeutics or cellular therapies would certainly benefit from an interdisciplinary approach, and one might say they even require it. Cell-based therapy parties include medical practitioners, patients, parties that engineer and manage a production pipeline, and scientists who engineer human cell lines or unicellular organisms [170]. Intercommunication between the parties and forming feedback mechanisms will optimize the flow of information and products under feasible conditions (fair price of a treatment, patient accessibility to the treatment, applicability of a treatment by medical practitioners and facilities, etc.) [171]. A feedback mechanism between patient and lab is especially important in personalized medicine and tailored treatments. Collaborative frameworks that help parties interact, share experience and data, and integrate new information into a cohesive understanding toward the overarching goal, while managing current challenges and refreshing side goals, tend to be underestimated. However, organized systems respond better than individuals to ongoing issues in the world [172].

The importance of interdisciplinary research organization has never been higher in the history of science. The challenge to solve century-old complex questions requires a new set of thinking. Organizations are at the heart of modern research approaches. Among these strategic organizations, research hubs are of exceptional importance. In a collection research direction that converges biology, medicine, and engineering, establishment of biohubs across geographically closer universities brought together researchers that have distinct scientific heritage [173, 174]. Chan Zuckerberg (CZ) Biohub is one such example that leads diverse interdisciplinary research programs dealing with complicated and multistep issues of utmost importance for the benefit of human health [175]. Each biohub is designed as a research institute. Besides that, CZ Biohub acts as a medium that will facilitate collaborations between researchers. The successful projects from researchers of Stanford, UCSF, and Berkeley are awarded grants and usage of biohub facilities to speed up their progress within a directed framework [176]. CZ Biohub also hires independent researchers to work in their institutes at San Francisco Bay Area, New York, and Chicago. Since biohub is an interdisciplinary research strategy with less than 10 years of history, the impact of this model is yet to have been observed.

Like CZ Biohub; interdisciplinary research committees, initiatives, and scientific organizations for live biotherapeutic products (LBPs) have been established in some US universities to govern current research directions into more meaningful and synergistic efforts. The first is living therapeutics initiative (LTI) by UC San Francisco, planning to manage a total investment of US$250 million [177]. This scheme covers all the stages of a typical therapeutic modality, replacing the classical small-molecule drugs or biopharmaceutical drugs with living biotherapeutic products (LBPs). UCSF projects within the scope of LTI use both living human and bacterial cell-derived therapeutics to address difficulties in treating diseases that are difficult to cure or even mitigate [178]. This initiative attracted a life sciences company, Thermo Fisher Scientific Inc., to expand the manufacturing capabilities of research groups and smaller companies by installing a manufacturing hub inside the UCSF campus as part of LTI [179]. Another initiative was commenced by Sarafan ChEM-Hub and Department of Bioengineering at Stanford University, called the "Microbiome Therapies Initiative (MITI)" [180]. This initiative is solely focused on microbial organisms as living therapeutic agents with gifts amounting to US$17 million donated by several philanthropists. The initiative that is directed by the institute scholar Michael Fischbach of Sarafan ChEM-H and bioengineering aims at forming interdisciplinary collaborations within Stanford University to produce candidate therapy modalities to enter early clinical stages. MITI projects are centered around two main themes, the first is to develop new methods to engineer gut microbiome species that are missing established genetic engineering protocols and the second is to formulate microbial concoctions that will function as a gut microbiome healer [181]. The Novo Nordisk Foundation has stepped up to initiate a facility in the Technical University of Denmark. The facility will be devoted to the cell therapy products that can be used in many subject areas of cell therapy research, such as Parkinson's, type-1 diabetes, and several forms of cancer. The facility is planned to have a structure that will fine-tune the candidate living cell therapy drugs before clinical trials begin [182].

Transformative technologies require extensive collaboration between companies and research groups. In the area of living biotherapeutics, many companies work side by side to achieve novel functionalities [183]. Living therapeutics is still in its infancy and requires monetary, facility, strategic, and commercial support from much bigger and older pharmaceutical companies [184]. This is still a challenge since global pharmaceutical companies generally rely on the production of chemical compounds and formulas that requires different set of production standards than a living drug. Living drugs are subjected to vast differences between batches; therefore, production pipeline should be designed to minimize batch-to-batch variability. Additionally, storage conditions allow a living culture to keep all its capabilities. Obviously, a living drug should stay alive and functional until it is administered to the patient. Development of living therapeutics is considered as an emerging discipline and the team that produces the treatment method should have a diverse set of skills, as well as the capacity to work in an environment of people with diverse backgrounds. In the end, any company or translational research group that is willing to produce living therapeutics should actively seek a parent company to test, scale up, and commercialize their treatment options [185]. This need sometimes becomes so significant and difficult to solve that frequently inexperienced living therapeutics companies require another company (called CDMO, contract development and manufacturing organization) to secure a contract with a pharmaceutical giant [186]. Consultancy provided by CDMOs is helpful in all the stages of drug commercialization, starting from the drug prototype to manufacturing pipelines until commercialization, that is why these companies are mostly experienced at these levels. CDMOs often provide facility for treatment development that is tailored for living therapeutics. CDMOs also help living therapeutics research groups in the rapidly changing regulatory environment, which happens mostly because of constant changes in necessities of this emerging approach for treatment. All these services result in faster progress for successful treatment modalities [187].

Although the scope and regulations vary, in the context of regulatory agencies in Europe (European Pharmacopeia – Ph. Eur.) and the USA (Food and Drug Administration – FDA), the common name for living therapeutics is called "live biotherapeutic products" [188, 189]. Gut microbiota is a complicated cocktail consisting of mostly harmless commensals and despite similarities between humans, composition and spatial distribution of microorganism community are unique for a person. This uniqueness is even enforced by the behavior of human host, personal decisions on diet and program of physical activities influence the microbiota of different locations of gut. The quality–quantity parameters and switching threshold for healthy microbiota (eubiosis) and imbalanced microbiota (dysbiosis) is not a well-defined concept, whereas microbiota of a diseased person that is radically different from a healthy person's microbiota is a very good indicator of what may need to be replaced to recalibrate gut microbiome balance [190]. Regulatory bodies regard microbiota balance and uniqueness as an intriguing factor and a problem in the case of a disease that needs to be addressed thoroughly [191]. Because of microbiota's inherent specificity between people, it only seems natural to assume microbiota of animals and humans differ at every level, and this would further reduce predictive power or safety assessments of animal studies and the success rate of any LBPs in clinical trials [192]. Animal studies still remain the best preclinical option to imitate gut microbiomes, until organ-on-a-chip models imitate human gut microbiomes and diseases perfectly. Risk-laden nature of LBPs requires careful planning and execution of strategic elements in a well-defined order. What makes LBPs different from probiotics is mainly their intended use to cure or mitigate diseases [193–195]. Probiotics, on the other hand, are accepted as a food or supplement that are used by healthy people for generations safely [196]. The major health risk associated with LBPs is the transfer of therapeutic agents into the bloodstream to induce extensive infectious response for people with compromised immune functionalities [197]. To adequately respond to all potential concerns, researchers team up with experienced companies.

A recent collaboration between Ferring, Rebiotix, and MyBiotics aims at developing a living therapeutics to combat the disease of bacterial vaginosis, which causes the rate of miscarriages to increase and infertility problems [198]. MyBiotics Ltd. is expected to develop technologies to rebalance vaginal microbiota, while Ferring and Rebiotix will focus on clinical level biotherapeutics studies. In 2023, Microbiotica started a clinical trial collaboration with Merck & Co., Inc. where they work toward using bacterial strain consortium (MB097) that will increase the efficacy of an immune checkpoint inhibitor (anti-PD-1), KEYTRUDA® [199, 200]. Seres Therapeutics has entered a collaboration with Bacthera, a company that is specialized in CMDO [201]. By being a joint venture between Lonza Group and Chr. Hansen, Bacthera will provide a commercial manufacturing facility for Seres to produce LBP called SER-109 commercially for the first time ever, drug is now called VOWST™. SER-109 contains Firmicutes spores purified from healthy people and was shown to be an effective oral therapy agent against recurring *C. difficile* infections where both strains compete for similar metabolites [202, 203]. In 2018, Robert S. Langer's Sigilon Therapeutics and Eli Lilly and Company announced a research collaboration that intends to develop insulin-producing cells that are encapsulated to treat chronic type 1 diabetes [204]. Afibromer™ encapsulation technology developed by Sigilon was used to protect iPSC cells that are differentiated into pancreatic beta-cells from foreign body reactions, this was the first cell therapy for diabetes and the product was later named SIG-002 [205–207]. While Sigilon focused on cellular engineering and delivery-stealth systems, Eli Lilly provided all the necessary funding, investment into its company, and global support at clinical trial stage and commercialization in the case of collaborative success. Eli Lilly later bought Sigilon Therapeutics [208]. Collaboration is not limited to the interactions of two companies. Finch Therapeutics Group initiated a clinical trial collaboration with Brigham and Women's Hospital to try their complete consortia microbiome drug (CP101) in UC, although this collaboration will be valid until the end of clinical trials [209, 210]. Gusto Global's GUT-108 was shown to be a potent candidate for the treatment of IBD in a study [211, 212], which led to a research collaboration with Crohn's and Colitis Foundation [213]. Another IBD living therapeutics is developed through Roche-Synlogic partnership [214]. Senti Bio company that was founded by the same MIT professors has started an extensive multiyear collaboration with Spark Therapeutics, which is a member of the Roche Group [215]. Senti Bio is a world leader in gene circuit design and implementation of gene therapy modalities for CNS, eye, or liver-related diseases. Their genetic logic gates will sense cellular biomarkers to compute and execute a dynamic response upon changing conditions [216, 217]. The same company signed another contract with a Celest Therapeutics, a Chinese company to develop Gene Circuit platform-based living therapeutic drug to treat solid tumors [218]. A partnership between Ginkgo Bioworks and Microba Life Sciences that allows Gingko to screen strain bank of Microba using their high-throughput has begun in 2022 [219]. Screening of large set of strains isolated by Microba from gut microbiome will reveal LBP candidates against autoimmune diseases such as lupus, arthritis, and autoimmune liver diseases [220, 221]. The promise of CAR-T cell therapy generated numerous collaborations. Recently, Umoja Biopharma and AbbVie have signed a deal to develop in situ CAR-T cell therapies for hematologic malignancies and other types of cancer [222]. Umoja's VivoVec™ propriety lentiviral gene delivery platform enables reduced time lags of CAR-T cell preparation, the necessity of ex vivo CAR-T manipulation and forgoing the requirement of lymphodepletion [223]. A similar T-cell-based therapy strategy was planned to put into effect by BlueRock Therapeutics (a subsidiary of Bayer AG) and bit.bio [224]. In this collaborative agreement, iPSC-derived regulatory T cells that may target a wide range of autoimmune diseases will be generated by opti-ox™ cell programming technology developed by bit.bio [225, 226].

Collaborative mindset is especially critical for developing or implementing novel gene therapy modalities in developing countries [227]. Living therapeutics market share is still a fraction of all types of possible medical treatments. Additionally, decision-makers are still unsure about the regulatory scope of living therapeutics. These complexities might make the conditions for a living therapeutics company product to emerge from a developing country significantly challenging. It is demanding even to implement gene therapies and many forms of living therapeutics in those areas of the world. NIH and Bill & Melinda Gates Foundation initiated a collaboration to treat HIV and sickle cell disease in 2019 that would last for 7–10 years to complement the bigger initiative proposed and funded by US government, ending the HIV epidemic in the United States [228]. Gates Foundation's plan includes introduction of antiretroviral gene therapy modalities that are globally available, low cost, and highly efficient for patients currently living or are from sub-Saharan Africa [229].

5.4.1 The role of bioinformatics and computational biology in living therapeutics

With the advent of artificial intelligence (AI) software and tools, AI/ML (machine learning) usage in one or more aspects of a research project has become a reality. There has never been a better time to start complex, multi-institute and interdisciplinary research initiative with the computer tools that are being developed and prospects for better and faster software tools are always evaluated optimistically [230, 231]. Combining transformative technology of AI with living therapeutics is one of the best approaches to deal with current challenges for treating complex diseases.

The recent concept of self-driving labs is one of the most promising approaches to deal with issues in any drug development method including living therapeutics [232]. It is a concept where ML, automation, and robotics cross-feed each other until a scientifically relevant output that is defined by the user is achieved [233, 234]. What happens in self-driving labs can be crudely explained in a classical lab setting where AI takes over the role of a principal investigator whereas robotics is the grad student workforce that performs the actual experiments. However, the decision for the next set of experiments is made almost instantly, thereby allowing large-scale data production intelligently. The robotic equipment for the transfer of solid and liquid material can be stationary, mobile, or fluidically through thin tubes with the help of a pump. AI software and tools that perform decision-making and commanding hardware can be based on mathematical modeling and analysis. AI can also tailor mathematical models with computer predictions based on calculations. AI-robotics-based detailed system control, precise actuation and tailored approach will strengthen the toolbox against inherent problems of batch-to-batch variability and standardization of product pipelines in living therapeutics. The success rate of candidate drugs can be expected to increase for the products that are optimized in self-driving labs [235].

The number, quality, robustness, and speed of AI tools that a researcher can use have exponentially increased within the last 5 years. ML-based computer models are being added to the classical bioinformatics computer program tools that are being heavily used in genetics and molecular biology [236]. Although AI/ML models are increasing very rapidly, their extent and applicability to the live biotherapeutics products are very limited. A significant portion of living therapeutics deals with the manipulation of concentration, types, and bioavailability of metabolites that are produced in the patient's body [237]. In the case of metabolic engineering, AI tools have been shown to be integrated into the decision-making process to promote target biomolecule production. Well-known ML methods like deep neural network and differential search algorithm were trained on molecular metabolomics data to decide which genes in a pathway can be silenced to achieve greater output for the target metabolite or transcription factor that control a set of genes [238, 239]. Biosensor-based training data has been utilized to train another Bayesian AI tools pair called "automatic recommendation tool" and EVOLVE algorithms to determine genetic constructs that need to be produced for boosting tryptophan production in *S. cerevisiae* [240]. In another study, the production of a chemotherapy drug paclitaxel in hazel shells through a collection of endophyte yeast strains was increased with the help of an artificial neural network algorithm [241]. Mathematical model prediction and its curation to improve its performance is also achieved by ML algorithms. AMMEDEUS tool can analyze multiple simulations that are generated by metabolic models to reannotate and improve the models through statistical learning [242]. ML tools can even be used to validate and analyze multi-omics data that is very difficult to assess with classical bioinformatics tools [243]. Tools based on AI/ML methods will become more relevant to the development of LBPs as the limits of classical bioinformatic tools are reached.

5.5 Challenges and limitations

Cellular engineering for living therapeutics poses a multifaceted challenge that demands. A variety of challenges must be overcome to fully realize the promise of modified cells for therapeutic applications in the field of cellular engineering for living therapies. Focusing on creating cellular structures with high cellular density and specialized functionalities, cellular engineering is developing quickly [244].

Although the human microbiome is an attractive living therapeutics modality, either through probiotics or reprogrammed bacteria, significant hurdles must be addressed. Undeniably, there is a lack of knowledge on host–microbiome interactions because of complex relations at multiple levels. It is expected that the state-of-the-art techniques in omics technologies will fill the gaps in the future. A similar problem arises from unculturable organisms of the microbiome. Most human microbiomes are understudied, but the gut microbiome could be an exception to some degree. In addition, the spatiotemporal regulation of the microbiome would be valuable in precisely intervening in an already complex environment. A ubiquitous problem is the genetic stability of the engineered bacteria. The metabolic burden and fitness pressure trigger adaptive mutations in the genome, which decreases or eliminates the expression of engineered genes. The high number of constructed genetic circuits inevitably increases the metabolic load on engineered bacteria, eventually increasing the risk of failure.

Moreover, the manufacturing process of microorganisms should be designed to fit the engineered mechanism's inner workings; for instance, facultative anaerobic bacteria need to be cultivated in a strictly controlled oxygen environment. Because genetically modified organisms could exchange genetic information with their wild-type counterparts, it is imperative to decrease the risk of engineered organisms' escape in nature, which could be achieved using kill switches or auxotrophic systems. A potentially robust solution is to use toxin–antitoxin systems found in some bacteria [245]. A remarkable strategy to avoid gene transfer from engineered cells to wild-type bacteria is to build a genetic firewall. Recorded genomes were immune to viral attacks and prevented bidirectional gene transfer because of their swapped genetic code [246, 247]. Another consideration for living therapeutics is monitoring the outcome through biomarkers of therapeutic activity. It is desirable that an easily measurable biomarker, ideally excreted in urine, could be produced only when the designed circuit is on. Pharmacokinetic and pharmacodynamic (PK/PD) tools are indispensable for drug development. Although it is more complex than living therapeutics, PK/PD tools are needed for engineered microorganisms. The route of delivery, real-time biodistribution data, and in vivo imaging all need to be established for translation from lab to clinic. It is worth noting that the US FDA released a guideline for clinical trials involving live therapeutic products [248]. As the clinical trials of engineered bacteria are being conducted, more complex setups will likely be required. A guiding principle for such cases could be previously engineered cell-based therapies that have been approved, such as CAR-T cells. Although there are serious challenges regarding efficacy, translation, manufacturing, and regulatory approvals, it is important to remember that live bacterial therapeutics have been used safely to save lives for decades, as in the case of Bacillus Calmette-Guérin, which is not only used as a tuberculosis vaccine but was also approved in the 1970s as an immunotherapeutic treatment for non-muscle-invasive bladder cancer [249]. There is no reason engineered bacteria cannot further enhance our microbiome-related therapeutic toolbox to address hard-to-treat diseases in the coming years.

Developing synthetic cell therapy procedures, evolving molecular devices for cellular control, and biofabricating organ-specific tissues are just a few approaches scientists are investigating to accomplish this goal [250, 251]. Additional hurdles for living pharmaceuticals include design, manufacturing efficiency, and quality assurance in synthetic cell treatments [251, 252]. Robust frameworks ensuring the quality and safety of cellular products intended for therapeutic use must be established. Furthermore, integrating artificial and natural cells to control cellular behavior and responses to external stimuli appears promising for advancing cellular engineering for therapeutic applications [253].

Prolonged administration of changed cellular material may jeopardize the safety and efficacy of the therapeutic approach because of potential immunogenicity, TCM transformation, and overgrowth [254]. Obstacles like poor lung compression, poor retention and engraftment, insufficient stability during storage and transportation, and safety concerns have prevented cell treatments from smoothly transferring into clinical applications [255]. The development of *Bacteroides* thetaiotaomicron as a live biotherapeutic platform is hampered by the scarcity of genetic components and technology, which may restrict the applicability and efficacy of genetic engineering methods in this context [256]. One of the existing drawbacks in cultured cellular networks is promiscuous connection. This clashes with the splendid fidelity of the body's inherent circuits regarding cell identity and connection specificity, which may affect the functionality and dependability of synthetic cellular networks [257]. Targeting proteins have been successfully inserted into cells to produce therapeutic effects. However, protein assembly is usually insufficient to manipulate the large-scale characteristics or architectures of cellular "machines" needed for disease modulation [258]. Altered bacterial strains must be thoroughly assessed during their preclinical development to facilitate their eventual clinical use. Developing more intricate, robust, and dependable synthetic bacteria for medical and diagnostic applications will require new techniques [259].

5.6 Future perspective

Significant advancements and revolutions are anticipated in the realm of living therapies in the future. Trends and predictions suggest that gut probiotics will be engineered using synthetic biology to treat specific diseases [260]. By using gut-engineered probiotics as live medicines, this method gives hope for treating diseases, including PKU and IBS. Furthermore, the potential for designed systems to sense and react to many contexts intelligently, enhancing specificity and efficacy beyond traditional therapies, is demonstrated by using programmable bacteria to promote tumor regression and systemic antitumor immunity [261]. The study conducted by Skylar-Scott et al. [244] emphasizes the need for innovative methods in biofabrication to overcome the challenges associated with producing tissues that are both therapeutically effective and can be reused. Researchers have established the way for future advancements in organ-specific tissue engineering by harnessing the distinctive characteristics of organoids generated from iPSCs using SWIFT techniques. This work contributed to the expanding body of research focused on optimizing cellular engineering for the development of advanced tissue constructs that have improved functions and therapeutic effectiveness. It provided valuable insights into the creation of next-generation living pharmaceuticals. Another study examined the modification of the molecular structure of the cell surface to enhance the therapeutic capabilities by modifying the glycocalyx, which is a protein and sugar layer present on the surface of all living cells. By meticulously altering the glycocalyx on living cells, this approach has the ability to alter cell behavior and enhance cell survival [262]. Li et al. [262] have shown that modifying the glycocalyx of living cells is a promising method for tailoring the surface properties of cells to meet specific therapeutic requirements [262]. Yang et al. [263] primarily focus on active shell cell nanobiohybrid systems in their discussion of using single-cell surface engineering to enhance therapeutic living cells through bioaugmentation. To enhance treatment outcomes, this prospective research avenue aims to develop effective communication between foreign entities present on cell surfaces and within cells. Imagine a future when advanced modifications to cell surfaces, along with biofabrication technologies, create living structures that can reproduce tissues or organs and have therapeutic capabilities [263]. The 2022 study conducted by Sarbadhikary and his colleagues explored the potential future applications of biophotonics, such as cellular optoelectronics and small live lasers, for advanced imaging, diagnostic, and therapeutic purposes. This advancement has the potential to create new opportunities for precision medicine by completely transforming the methods of cellular imaging and therapy [264]. The future of living therapeutics is promising, with ongoing advances in gene editing, synthetic biology, and personalized medicine. As the field continues to evolve, interdisciplinary collaboration, technological innovation, and ethical stewardship will be crucial in bringing these groundbreaking therapies to patients worldwide. By addressing the challenges and embracing the opportunities, living therapeutics can revolutionize healthcare and improve the lives of millions of people.

5.7 Conclusion

The latest developments in cellular engineering represent a notable milestone in medical therapies, introducing a new age characterized by accuracy, effectiveness, and selectivity. This chapter elucidates the transition from conventional therapies to dynamic therapeutic systems, emphasizing the groundbreaking capabilities of synthetic biology. Scientists have created autonomous systems that can identify, cure, and potentially eradicate different illnesses by utilizing advanced genetic circuits. Live treatments have numerous benefits compared to conventional medications, including precise treatment using modified microorganisms, reduced manufacturing expenses, and a decreased occurrence of adverse reactions.

Investigating the human microbiota as a therapeutic platform offers hopeful prospects for utilizing live therapeutics. The correlation between the microbiome and human health underscores the capacity of microbiome engineering to rectify equilibrium and tackle various illnesses. Scientists have the ability to alter the makeup and operation of microbial populations in the GI, dermal, respiratory, and urogenital systems to address conditions such as obesity, COPD, AD, and IBD. Utilizing probiotics, in conjunction with prebiotics and postbiotics, is a pragmatic method to reinstate equilibrium in the microbiome and augment therapeutic results. Efforts to address the pressing problem of AMR are underway through the use of cutting-edge treatments, including the application of engineered microorganisms and phages. Phages provide a focused and efficient alternative to broad-spectrum antibiotics due to their ability to selectively target and eliminate specific bacterial species. CRISPR-Cas9 technology improves the efficacy of phage therapy by facilitating accurate genetic alterations that selectively eradicate dangerous germs while safeguarding good ones. In addition, the utilization of bacterial secretion systems and regulated microbial consortia offers a versatile method for addressing resistant illnesses and enhancing general health.

The use of contemporary cellular engineering technology in medical treatments shows immense potential for the future of medicine. Researchers are utilizing synthetic biology concepts to create sophisticated living medicines that can precisely and efficiently target medical requirements that were previously unaddressed. The improvements discussed in this chapter establish the groundwork for a future in which personalized, accurate, and self-regulating treatments become the standard, leading to substantial enhancements in patient results and revolutionizing the healthcare sector. To properly utilize these technological developments and assure their extensive application in clinical environments, it is imperative to persist in research efforts, foster interdisciplinary cooperation, and carry out clinical studies. As technology advances, the integration of biology and engineering will surely result in revolutionary therapeutic methods that rethink our approach to treating and curing diseases.

References

[1]   Cubillos-Ruiz A, Guo T, Sokolovska A, Miller PF, Collins JJ, Lu TK, et al. Engineering living therapeutics with synthetic biology. Nat Rev Drug Discov 2021;20:941–60. https://doi.org/10.1038/s41573-021-00285-3.

[2]   Fischbach MA, Bluestone JA, Lim WA. Cell-based therapeutics: The next pillar of medicine. Sci Transl Med 2013;5:179ps7–179ps7. https://doi.org/10.1126/scitranslmed.3005568.

[3]   Liu X, Wu M, Wang M, Duan Y, Phan C, Qi G, et al. Metabolically engineered bacteria as light-controlled living therapeutics for anti-angiogenesis tumor therapy. Mater Horiz 2021;8:1454–60. https://doi.org/10.1039/D0MH01582B.

[4]   Pedrolli DB, Ribeiro NV, Squizato PN, De jesus VN, Cozetto DA. Team AQA Unesp at iGEM 2017. Engineering microbial living therapeutics: The synthetic biology toolbox. Trends Biotechnol 2019;37:100–15. https://doi.org/10.1016/j.tibtech.2018.09.005.

[5]   Nandagopal N, Elowitz MB. Synthetic biology: Integrated gene circuits. Science 2011;333:1244–48. https://doi.org/10.1126/science.1207084.

[6]   Kitada T, DiAndreth B, Teague B, Weiss R. Programming gene and engineered-cell therapies with synthetic biology. Science 2018;359:eaad1067. https://doi.org/10.1126/science.aad1067.

[7]   Adams BL. The next generation of synthetic biology chassis: Moving synthetic biology from the laboratory to the field. ACS Synth Biol 2016;5:1328–30. https://doi.org/10.1021/acssynbio.6b00256.

[8]   Mimee M, Tucker AC, Voigt CA, Lu TK. Programming a human commensal bacterium, bacteroides thetaiotaomicron, to sense and respond to stimuli in the murine gut microbiota. Cell Syst 2015;1:62–71. https://doi.org/10.1016/j.cels.2015.06.001.

[9]   Pantoja Angles A, Ali Z, Mahfouz M. CS-Cells: A CRISPR-Cas12 DNA device to generate chromosome-shredded cells for efficient and safe molecular biomanufacturing. ACS Synth Biol 2022;11:430–40. https://doi.org/10.1021/acssynbio.1c00516.

[10] Pöhlmann C, Brandt M, Mottok DS, Zschüttig A, Campbell JW, Blattner FR, et al. Periplasmic delivery of biologically active human interleukin-10 in Escherichia coli via a sec-dependent signal peptide. J Mol Microbiol Biotechnol 2012;22:1–9. https://doi.org/10.1159/000336043.

[11] Förster S, Brandt M, Mottok DS, Zschüttig A, Zimmermann K, Blattner FR, et al. Secretory expression of biologically active human Herpes virus interleukin-10 analogues in Escherichia colivia a modified Sec-dependent transporter construct. BMC Biotechnol 2013;13:82. https://doi.org/10.1186/1472-6750-13-82.

[12] Turnbaugh PJ, Ley RE, Hamady M, Fraser-Liggett CM, Knight R, Gordon JI. The human microbiome project. Nature 2007;449:804–10. https://doi.org/10.1038/nature06244.

[13] Mackowiak PA. Recycling Metchnikoff: Probiotics, the intestinal microbiome and the quest for long life. Front Public Health 2013;1. https://doi.org/10.3389/fpubh.2013.00052.

[14] Hacker J, Blum-Oehler G. In appreciation of theodor escherich. Nat Rev Microbiol 2007;5:902–902. https://doi.org/10.1038/nrmicro1810.

[15] Prescott SL. History of medicine: Origin of the term microbiome and why it matters. Hum Microbiome J 2017;4:24–25. https://doi.org/10.1016/j.humic.2017.05.004.

[16] Baquero F, Nombela C. The microbiome as a human organ. Clin Microbiol Infect 2012;18:2–4. https://doi.org/10.1111/j.1469-0691.2012.03916.x.

[17] Berg G, Rybakova D, Fischer D, Cernava T, Vergès M-CC, Charles T, et al. Microbiome definition revisited: Old concepts and new challenges. Microbiome 2020;8:103. https://doi.org/10.1186/s40168-020-00875-0.

[18] Hrncir T. Gut microbiota dysbiosis: Triggers, consequences, diagnostic and therapeutic options. Microorganisms 2022;10:578. https://doi.org/10.3390/microorganisms10030578.

[19] Aggarwal N, Kitano S, Puah GRY, Kittelmann S, Hwang IY, Chang MW. Microbiome and human health: Current understanding, engineering, and enabling technologies. Chem Rev 2023;123:31–72. https://doi.org/10.1021/acs.chemrev.2c00431.

[20] Hanssen NMJ, De Vos WM, Nieuwdorp M. Fecal microbiota transplantation in human metabolic diseases: From a murky past to a bright future? Cell Metab 2021;33:1098–110. https://doi.org/10.1016/j.cmet.2021.05.005.

[21] Kim KO, Gluck M. Fecal microbiota transplantation: An update on clinical practice. Clin Endosc 2019;52:137–43. https://doi.org/10.5946/ce.2019.009.

[22] Walker AW, Hoyles L. Human microbiome myths and misconceptions. Nat Microbiol 2023;8:1392–96. https://doi.org/10.1038/s41564-023-01426-7.

[23] Hatton IA, Galbraith ED, Merleau NSC, Miettinen TP, Smith BM, Shander JA. The human cell count and size distribution. Proc Natl Acad Sci 2023;120:e2303077120. https://doi.org/10.1073/pnas.2303077120.

[24] Sender R, Fuchs S, Milo R. Are we really vastly outnumbered? Revisiting the ratio of bacterial to host cells in humans. Cell 2016;164:337–40. https://doi.org/10.1016/j.cell.2016.01.013.

[25] Bischoff SC, Barbara G, Buurman W, Ockhuizen T, Schulzke J-D, Serino M, et al. Intestinal permeability – a new target for disease prevention and therapy. BMC Gastroenterol 2014;14:189. https://doi.org/10.1186/s12876-014-0189-7.

[26] Stolfi C, Maresca C, Monteleone G, Laudisi F. Implication of intestinal barrier dysfunction in gut dysbiosis and diseases. Biomedicines 2022;10:289. https://doi.org/10.3390/biomedicines10020289.

[27] Sankarasubramanian J, Ahmad R, Avuthu N, Singh AB, Guda C. Gut microbiota and metabolic specificity in ulcerative colitis and Crohn's disease. Front Med 2020;7:606298. https://doi.org/10.3389/fmed.2020.606298.

[28] Lopez-Siles M, Duncan SH, Garcia-Gil LJ, Martinez-Medina M. *Faecalibacterium* prausnitzii : From microbiology to diagnostics and prognostics. ISME J 2017;11:841–52. https://doi.org/10.1038/ismej.2016.176.

[29] Sokol H, Pigneur B, Watterlot L, Lakhdari O, Bermúdez-Humarán LG, Gratadoux -J-J, et al. *Faecalibacterium* prausnitzii is an anti-inflammatory commensal bacterium identified by gut microbiota analysis of Crohn disease patients. Proc Natl Acad Sci 2008;105:16731–36. https://doi.org/10.1073/pnas.0804812105.

[30] Chong PP, Chin VK, Looi CY, Wong WF, Madhavan P, Yong VC. The microbiome and irritable Bowel syndrome – A review on the pathophysiology, current research and future therapy. Front Microbiol 2019;10:1136. https://doi.org/10.3389/fmicb.2019.01136.

[31] Valitutti C, Fasano. Celiac disease and the microbiome. Nutrients 2019;11:2403. https://doi.org/10.3390/nu11102403.

[32] Wong CC, Yu J. Gut microbiota in colorectal cancer development and therapy. Nat Rev Clin Oncol 2023;20:429–52. https://doi.org/10.1038/s41571-023-00766-x.

[33] Carco C, Young W, Gearry RB, Talley NJ, McNabb WC, Roy NC. Increasing evidence that irritable bowel syndrome and functional gastrointestinal disorders have a microbial pathogenesis. Front Cell Infect Microbiol 2020;10:468. https://doi.org/10.3389/fcimb.2020.00468.

[34] Barbara G, Barbaro MR, Fuschi D, Palombo M, Falangone F, Cremon C, et al. Inflammatory and microbiota-related regulation of the intestinal epithelial barrier. Front Nutr 2021;8:718356. https://doi.org/10.3389/fnut.2021.718356.

[35] Psaltis AJ, Mackenzie BW, Cope EK, Ramakrishnan VR. Unraveling the role of the microbiome in chronic rhinosinusitis. J Allergy Clin Immunol 2022;149:1513–21. https://doi.org/10.1016/j.jaci.2022.02.022.

[36] Ramsheh MY, Haldar K, Esteve-Codina A, Purser LF, Richardson M, Müller-Quernheim J, et al. Lung microbiome composition and bronchial epithelial gene expression in patients with COPD versus healthy individuals: a bacterial 16S rRNA gene sequencing and host transcriptomic analysis. Lancet Microbe 2021;2:e300–10. https://doi.org/10.1016/S2666-5247(21)00035-5.

[37] Frati F, Salvatori C, Incorvaia C, Bellucci A, Di Cara G, Marcucci F, et al. The role of the microbiome in asthma: The gut–lung axis. Int J Mol Sci 2018;20:123. https://doi.org/10.3390/ijms20010123.

[38] Martin CR, Osadchiy V, Kalani A, Mayer EA. The brain-gut-microbiome axis. Cell Mol Gastroenterol Hepatol 2018;6:133–48. https://doi.org/10.1016/j.jcmgh.2018.04.003.

[39] Oh J, Byrd AL, Park M, Kong HH, Segre JA. Temporal stability of the human skin microbiome. Cell 2016;165:854–66. https://doi.org/10.1016/j.cell.2016.04.008.

[40] Feng H, Shuda M, Chang Y, Moore PS. Clonal integration of a polyomavirus in human merkel cell carcinoma. Science 2008;319:1096–100. https://doi.org/10.1126/science.1152586.

[41] McLaughlin J, Watterson S, Layton AM, Bjourson AJ, Barnard E, McDowell A. Propionibacterium acnes and acne vulgaris: New insights from the integration of population genetic, multi-omic, biochemical and host-microbe studies. Microorganisms 2019;7:128. https://doi.org/10.3390/microorganisms7050128.

[42] Koh LF, Ong RY, Common JE. Skin microbiome of atopic dermatitis. Allergol Int 2022;71:31–39. https://doi.org/10.1016/j.alit.2021.11.001.

[43] Nakatsuji T, Gallo RL. The role of the skin microbiome in atopic dermatitis. Ann Allergy Asthma Immunol 2019;122:263–69. https://doi.org/10.1016/j.anai.2018.12.003.

[44] Myles IA, Castillo CR, Barbian KD, Kanakabandi K, Virtaneva K, Fitzmeyer E, et al. Therapeutic responses to *Roseomonas* mucosa in atopic dermatitis may involve lipid-mediated TNF-related epithelial repair. Sci Transl Med 2020;12:eaaz8631. https://doi.org/10.1126/scitranslmed.aaz8631.

[45] Ackerman AL, Chai TC. The bladder is not sterile: An update on the urinary microbiome. Curr Bladder Dysfunct Rep 2019;14:331–41. https://doi.org/10.1007/s11884-019-00543-6.

[46] Jones J, Murphy CP, Sleator RD, Culligan EP. The urobiome, urinary tract infections, and the need for alternative therapeutics. Microb Pathog 2021;161:105295. https://doi.org/10.1016/j.micpath.2021.105295.

[47] Zandbergen LE, Halverson T, Brons JK, Wolfe AJ, De Vos MGJ. The good and the bad: Ecological interaction measurements between the urinary microbiota and uropathogens. Front Microbiol 2021;12:659450. https://doi.org/10.3389/fmicb.2021.659450.

[48] Roth RS, Liden M, Huttner A. The urobiome in men and women: A clinical review. Clin Microbiol Infect 2023;29:1242–48. https://doi.org/10.1016/j.cmi.2022.08.010.

[49] Hilt EE, McKinley K, Pearce MM, Rosenfeld AB, Zilliox MJ, Mueller ER, et al. Urine is not sterile: Use of enhanced urine culture techniques to detect resident bacterial flora in the adult female bladder. J Clin Microbiol 2014;52:871–76. https://doi.org/10.1128/JCM.02876-13.

[50] Karstens L, Siddiqui NY, Zaza T, Barstad A, Amundsen CL, Sysoeva TA. Benchmarking DNA isolation kits used in analyses of the urinary microbiome. Sci Rep 2021;11:6186. https://doi.org/10.1038/s41598-021-85482-1.

[51] Neugent ML, Hulyalkar NV, Nguyen VH, Zimmern PE, De Nisco NJ. Advances in understanding the human urinary microbiome and its potential role in urinary tract infection. mBio 2020;11:e00218–20. https://doi.org/10.1128/mBio.00218-20.

[52] Thomas-White K, Brady M, Wolfe AJ, Mueller ER. The bladder is not sterile: History and current discoveries on the urinary microbiome. Curr Bladder Dysfunct Rep 2016;11:18–24. https://doi.org/10.1007/s11884-016-0345-8.

[53] Sonnenborn U. *Escherichia* coli strain Nissle 1917 – from bench to bedside and back: history of a special *Escherichia* coli strain with probiotic properties. FEMS Microbiol Lett 2016;363:fnw212. https://doi.org/10.1093/femsle/fnw212.

[54] Zhao Z, Xu S, Zhang W, Wu D, Yang G. Probiotic *Escherichia* coli NISSLE 1917 for inflammatory bowel disease applications. Food Funct 2022;13:5914–24. https://doi.org/10.1039/D2FO00226D.

[55] Khalesi S, Bellissimo N, Vandelanotte C, Williams S, Stanley D, Irwin C. A review of probiotic supplementation in healthy adults: helpful or hype? Eur J Clin Nutr 2019;73:24–37. https://doi.org/10.1038/s41430-018-0135-9.

[56] Borgeraas H, Johnson LK, Skattebu J, Hertel JK, Hjelmesæth J. Effects of probiotics on body weight, body mass index, fat mass and fat percentage in subjects with overweight or obesity: a systematic review and meta-analysis of randomized controlled trials. Obes Rev 2018;19:219–32. https://doi.org/10.1111/obr.12626.

[57] Depommier C, Everard A, Druart C, Plovier H, Van Hul M, Vieira-Silva S, et al. Supplementation with Akkermansia muciniphila in overweight and obese human volunteers: a proof-of-concept exploratory study. Nat Med 2019;25:1096–103. https://doi.org/10.1038/s41591-019-0495-2.

[58] Zhao H, Xu H, Chen S, He J, Zhou Y, Nie Y. Systematic review and meta-analysis of the role of *Faecalibacterium* prausnitzii alteration in inflammatory bowel disease. J Gastroenterol Hepatol 2021;36:320–28. https://doi.org/10.1111/jgh.15222.

[59] Wastyk HC, Perelman D, Topf M, Fragiadakis GK, Robinson JL, Sonnenburg JL, et al. Randomized controlled trial demonstrates response to a probiotic intervention for metabolic syndrome that may correspond to diet. Gut Microbes 2023;15:2178794. https://doi.org/10.1080/19490976.2023.2178794.

[60] Kothari D, Patel S, Kim S-K. Probiotic supplements might not be universally-effective and safe: A review. Biomed Pharmacother 2019;111:537–47. https://doi.org/10.1016/j.biopha.2018.12.104.

[61] Suez J, Zmora N, Zilberman-Schapira G, Mor U, Dori-Bachash M, Bashiardes S, et al. Post-antibiotic gut mucosal microbiome reconstitution is impaired by probiotics and improved by autologous FMT. Cell 2018;174:1406–23, e16. https://doi.org/10.1016/j.cell.2018.08.047.

[62] Wastyk HC, Fragiadakis GK, Perelman D, Dahan D, Merrill BD, Yu FB, et al. Gut-microbiota-targeted diets modulate human immune status. Cell 2021;184:4137–53, e14. https://doi.org/10.1016/j.cell.2021.06.019.

[63] Gopalakrishnan V, Spencer CN, Nezi L, Reuben A, Andrews MC, Karpinets TV, et al. Gut microbiome modulates response to anti–PD-1 immunotherapy in melanoma patients. Science 2018;359:97–103. https://doi.org/10.1126/science.aan4236.

[64] Matson V, Fessler J, Bao R, Chongsuwat T, Zha Y, Alegre M-L, et al. The commensal microbiome is associated with anti–PD-1 efficacy in metastatic melanoma patients. Science 2018;359:104–08. https://doi.org/10.1126/science.aao3290.

[65] Routy B, Le Chatelier E, Derosa L, Duong CPM, Alou MT, Daillère R, et al. Gut microbiome influences efficacy of PD-1–based immunotherapy against epithelial tumors. Science 2018;359:91–97. https://doi.org/10.1126/science.aan3706.

[66] Li X, Zhang S, Guo G, Han J, Yu J. Gut microbiome in modulating immune checkpoint inhibitors. eBioMedicine 2022;82:104163. https://doi.org/10.1016/j.ebiom.2022.104163.

[67] Zhang M, Liu J, Xia Q. Role of gut microbiome in cancer immunotherapy: from predictive biomarker to therapeutic target. Exp Hematol Oncol 2023;12:84. https://doi.org/10.1186/s40164-023-00442-x.

[68] Burnham C-AD, Leeds J, Nordmann P, O'Grady J, Patel J. Diagnosing antimicrobial resistance. Nat Rev Microbiol 2017;15:697–703. https://doi.org/10.1038/nrmicro.2017.103.

[69] Micoli F, Bagnoli F, Rappuoli R, Serruto D. The role of vaccines in combatting antimicrobial resistance. Nat Rev Microbiol 2021;19:287–302. https://doi.org/10.1038/s41579-020-00506-3.

[70] Larson MH, Gilbert LA, Wang X, Lim WA, Weissman JS, Qi LS. CRISPR interference (CRISPRi) for sequence-specific control of gene expression. Nat Protoc 2013;8:2180–96. https://doi.org/10.1038/nprot.2013.132.

[71] Darby EM, Trampari E, Siasat P, Gaya MS, Alav I, Webber MA, et al. Molecular mechanisms of antibiotic resistance revisited. Nat Rev Microbiol 2023;21:280–95. https://doi.org/10.1038/s41579-022-00820-y.

[72] The promise of phages. Nat Biotechnol 2023;41:583–583. https://doi.org/10.1038/s41587-023-01807-7.

[73] Czaplewski L, Bax R, Clokie M, Dawson M, Fairhead H, Fischetti VA, et al. Alternatives to antibiotics – a pipeline portfolio review. Lancet Infect Dis 2016;16:239–51. https://doi.org/10.1016/S1473-3099(15)00466-1.

[74] Strathdee SA, Hatfull GF, Mutalik VK, Schooley RT. Phage therapy: From biological mechanisms to future directions. Cell 2023;186:17–31. https://doi.org/10.1016/j.cell.2022.11.017.

[75] Jinek M, Chylinski K, Fonfara I, Hauer M, Doudna JA, Charpentier E. A programmable dual-RNA–guided DNA endonuclease in adaptive bacterial immunity. Science 2012;337:816–21. https://doi.org/10.1126/science.1225829.

[76] Citorik RJ, Mimee M, Lu TK. Sequence-specific antimicrobials using efficiently delivered RNA-guided nucleases. Nat Biotechnol 2014;32:1141–45. https://doi.org/10.1038/nbt.3011.

[77] Blasey N, Rehrmann D, Riebisch AK, Mühlen S. Targeting bacterial pathogenesis by inhibiting virulence-associated Type III and Type IV secretion systems. Front Cell Infect Microbiol 2022;12:1065561. https://doi.org/10.3389/fcimb.2022.1065561.

[78] Kreitz J, Friedrich MJ, Guru A, Lash B, Saito M, Macrae RK, et al. Programmable protein delivery with a bacterial contractile injection system. Nature 2023;616:357–64. https://doi.org/10.1038/s41586-023-05870-7.

[79] Ting S-Y, Martínez-García E, Huang S, Bertolli SK, Kelly KA, Cutler KJ, et al. Targeted depletion of bacteria from mixed populations by programmable adhesion with antagonistic competitor cells. Cell Host Microbe 2020;28:313–21, e6. https://doi.org/10.1016/j.chom.2020.05.006.

[80] Rubin BE, Diamond S, Cress BF, Crits-Christoph A, Lou YC, Borges AL, et al. Species- and site-specific genome editing in complex bacterial communities. Nat Microbiol 2022;7:34–47. https://doi.org/10.1038/s41564-021-01014-7.

[81] Lam KN, Spanogiannopoulos P, Soto-Perez P, Alexander M, Nalley MJ, Bisanz JE, et al. Phage-delivered CRISPR-Cas9 for strain-specific depletion and genomic deletions in the gut microbiome. Cell Rep 2021;37:109930. https://doi.org/10.1016/j.celrep.2021.109930.

[82] Selle K, Fletcher JR, Tuson H, Schmitt DS, McMillan L, Vridhambal GS, et al. *In* Vivo Targeting of Clostridioides difficile Using Phage-Delivered CRISPR-Cas3 Antimicrobials. mBio 2020;11:e00019–20. https://doi.org/10.1128/mBio.00019-20.

[83] Peery AF, Kelly CR, Kao D, Vaughn BP, Lebwohl B, Singh S, et al. AGA clinical practice guideline on fecal microbiota-based therapies for select gastrointestinal diseases. Gastroenterology 2024;166:409–34. https://doi.org/10.1053/j.gastro.2024.01.008.

[84] Yadegar A, Bar-Yoseph H, Monaghan TM, Pakpour S, Severino A, Kuijper EJ, et al. Fecal microbiota transplantation: Current challenges and future landscapes. Clin Microbiol Rev 2024;e00060–22. https://doi.org/10.1128/cmr.00060-22.

[85] Shkoporov AN, Hill C. Bacteriophages of the human gut: The "Known Unknown" of the microbiome. Cell Host Microbe 2019;25:195–209. https://doi.org/10.1016/j.chom.2019.01.017.

[86] Ianiro G, Mullish BH, Kelly CR, Sokol H, Kassam Z, Ng SC, et al. Screening of faecal microbiota transplant donors during the COVID-19 outbreak: Suggestions for urgent updates from an international expert panel. Lancet Gastroenterol Hepatol 2020;5:430–32. https://doi.org/10.1016/S2468-1253(20)30082-0.

[87] Khanna S, Sims M, Louie TJ, Fischer M, LaPlante K, Allegretti J, et al. SER-109: An oral investigational microbiome therapeutic for patients with recurrent clostridioides difficile Infection (rCDI). Antibiot Basel Switz 2022;11:1234. https://doi.org/10.3390/antibiotics11091234.

[88] Deter HS, Lu T. Engineering microbial consortia with rationally designed cellular interactions. Curr Opin Biotechnol 2022;76:102730. https://doi.org/10.1016/j.copbio.2022.102730.

[89] Jia X, Liu C, Song H, Ding M, Du J, Ma Q, et al. Design, analysis and application of synthetic microbial consortia. Synth Syst Biotechnol 2016;1:109–17. https://doi.org/10.1016/j.synbio.2016.02.001.

[90] van der Lelie D, Oka A, Taghavi S, Umeno J, Fan T-J, Merrell KE, et al. Rationally designed bacterial consortia to treat chronic immune-mediated colitis and restore intestinal homeostasis. Nat Commun 2021;12:3105. https://doi.org/10.1038/s41467-021-23460-x.

[91] Hahn J, Ding S, Im J, Harimoto T, Leong KW, Danino T. Bacterial therapies at the interface of synthetic biology and nanomedicine. Nat Rev Bioeng 2023;2:120–35. https://doi.org/10.1038/s44222-023-00119-4.

[92] Cubillos-Ruiz A, Alcantar MA, Donghia NM, Cárdenas P, Avila-Pacheco J, Collins JJ. An engineered live biotherapeutic for the prevention of antibiotic-induced dysbiosis. Nat Biomed Eng 2022;6:910–21. https://doi.org/10.1038/s41551-022-00871-9.

[93] Gurbatri CR, Arpaia N, Danino T. Engineering bacteria as interactive cancer therapies. Science 2022;378:858–64. https://doi.org/10.1126/science.add9667.

[94] Yu Y, Lin S, Chen Z, Qin B, He Z, Cheng M, et al. Bacteria-driven bio-therapy: From fundamental studies to clinical trials. Nano Today 2023;48:101731. https://doi.org/10.1016/j.nantod.2022.101731.

[95] Kurtz CB, Millet YA, Puurunen MK, Perreault M, Charbonneau MR, Isabella VM, et al. An engineered E. coli Nissle improves hyperammonemia and survival in mice and shows dose-dependent exposure in healthy humans. Sci Transl Med 2019;11:eaau7975. https://doi.org/10.1126/scitranslmed.aau7975.

[96] Isabella VM, Ha BN, Castillo MJ, Lubkowicz DJ, Rowe SE, Millet YA, et al. Development of a synthetic live bacterial therapeutic for the human metabolic disease phenylketonuria. Nat Biotechnol 2018;36:857–64. https://doi.org/10.1038/nbt.4222.

[97] Adolfsen KJ, Callihan I, Monahan CE, Greisen PJ, Spoonamore J, Momin M, et al. Improvement of a synthetic live bacterial therapeutic for phenylketonuria with biosensor-enabled enzyme engineering. Nat Commun 2021;12:6215. https://doi.org/10.1038/s41467-021-26524-0.

[98] Ramachandran G, Bikard D. Editing the microbiome the CRISPR way. Philos Trans R Soc Lond B Biol Sci 2019;374:20180103. https://doi.org/10.1098/rstb.2018.0103.

[99] Doudna JA. The promise and challenge of therapeutic genome editing. Nature 2020;578:229–36. https://doi.org/10.1038/s41586-020-1978-5.

[100] Peters JM, Colavin A, Shi H, Czarny TL, Larson MH, Wong S, et al. A comprehensive, CRISPR-based functional analysis of essential genes in bacteria. Cell 2016;165:1493–506. https://doi.org/10.1016/j.cell.2016.05.003.

[101] Groussin M, Poyet M, Sistiaga A, Kearney SM, Moniz K, Noel M, et al. Elevated rates of horizontal gene transfer in the industrialized human microbiome. Cell 2021;184:2053–67, e18. https://doi.org/10.1016/j.cell.2021.02.052.

[102] Ronda C, Chen SP, Cabral V, Yaung SJ, Wang HH. Metagenomic engineering of the mammalian gut microbiome in situ. Nat Methods 2019;16:167–70. https://doi.org/10.1038/s41592-018-0301-y.

[103] Mao N, Cubillos-Ruiz A, Cameron DE, Collins JJ. Probiotic strains detect and suppress cholera in mice. Sci Transl Med 2018;10:eaao2586. https://doi.org/10.1126/scitranslmed.aao2586.

[104] Kotula JW, Kerns SJ, Shaket LA, Siraj L, Collins JJ, Way JC, et al. Programmable bacteria detect and record an environmental signal in the mammalian gut. Proc Natl Acad Sci U S A 2014;111:4838–43. https://doi.org/10.1073/pnas.1321321111.

[105] Riglar DT, Giessen TW, Baym M, Kerns SJ, Niederhuber MJ, Bronson RT, et al. Engineered bacteria can function in the mammalian gut long-term as live diagnostics of inflammation. Nat Biotechnol 2017;35:653–58. https://doi.org/10.1038/nbt.3879.

[106] Coley WB. The treatment of malignant tumors by repeated inoculations of erysipelas. With a report of ten original cases. 1893. Clin Orthop 1991;3–11.

[107] Coley WB. The treatment of inoperable sarcoma by bacterial toxins (the mixed toxins of the streptococcus erysipelas and the bacillus prodigiosus). Proc R Soc Med 1910;3:1–48.

[108] Nejman D, Livyatan I, Fuks G, Gavert N, Zwang Y, Geller LT, et al. The human tumor microbiome is composed of tumor type-specific intracellular bacteria. Science 2020;368:973–80. https://doi.org/10.1126/science.aay9189.

[109] Din MO, Danino T, Prindle A, Skalak M, Selimkhanov J, Allen K, et al. Synchronized cycles of bacterial lysis for in vivo delivery. Nature 2016;536:81–85. https://doi.org/10.1038/nature18930.

[110] Chowdhury S, Castro S, Coker C, Hinchliffe TE, Arpaia N, Danino T. Programmable bacteria induce durable tumor regression and systemic antitumor immunity. Nat Med 2019;25:1057–63. https://doi.org/10.1038/s41591-019-0498-z.

[111] Gurbatri CR, Lia I, Vincent R, Coker C, Castro S, Treuting PM, et al. Engineered probiotics for local tumor delivery of checkpoint blockade nanobodies. Sci Transl Med 2020;12:eaax0876. https://doi.org/10.1126/scitranslmed.aax0876.

[112] Canale FP, Basso C, Antonini G, Perotti M, Li N, Sokolovska A, et al. Metabolic modulation of tumours with engineered bacteria for immunotherapy. Nature 2021;598:662–66. https://doi.org/10.1038/s41586-021-04003-2.

[113] Cooper RM, Wright JA, Ng JQ, Goyne JM, Suzuki N, Lee YK, et al. Engineered bacteria detect tumor DNA. Science 2023;381:682–86. https://doi.org/10.1126/science.adf3974.

[114] Luo H, Chen Y, Kuang X, Wang X, Yang F, Cao Z, et al. Chemical reaction-mediated covalent localization of bacteria. Nat Commun 2022;13:7808. https://doi.org/10.1038/s41467-022-35579-6.

[115] Harimoto T, Hahn J, Chen -Y-Y, Im J, Zhang J, Hou N, et al. A programmable encapsulation system improves delivery of therapeutic bacteria in mice. Nat Biotechnol 2022;40:1259–69. https://doi.org/10.1038/s41587-022-01244-y.

[116] Wang X, Lin S, Wang L, Cao Z, Zhang M, Zhang Y, et al. Versatility of bacterial outer membrane vesicles in regulating intestinal homeostasis. Sci Adv 2023;9:eade5079. https://doi.org/10.1126/sciadv.ade5079.

[117] Kim OY, Park HT, Dinh NTH, Choi SJ, Lee J, Kim JH, et al. Bacterial outer membrane vesicles suppress tumor by interferon-γ-mediated antitumor response. Nat Commun 2017;8:626. https://doi.org/10.1038/s41467-017-00729-8.

[118] Yue Y, Xu J, Li Y, Cheng K, Feng Q, Ma X, et al. Antigen-bearing outer membrane vesicles as tumour vaccines produced in situ by ingested genetically engineered bacteria. Nat Biomed Eng 2022;6:898–909. https://doi.org/10.1038/s41551-022-00886-2.

[119] Ali MK, Liu Q, Liang K, Li P, Kong Q. Bacteria-derived minicells for cancer therapy. Cancer Lett 2020;491:11–21. https://doi.org/10.1016/j.canlet.2020.07.024.

[120] Rampley CPN, Davison PA, Qian P, Preston GM, Hunter CN, Thompson IP, et al. Development of SimCells as a novel chassis for functional biosensors. Sci Rep 2017;7:7261. https://doi.org/10.1038/s41598-017-07391-6.

[121] Zhang Y, Ji W, He L, Chen Y, Ding X, Sun Y, et al. E. coli Nissle 1917-derived minicells for targeted delivery of chemotherapeutic drug to hypoxic regions for cancer therapy. Theranostics 2018;8:1690–705. https://doi.org/10.7150/thno.21575.

[122] Lim B, Yin Y, Ye H, Cui Z, Papachristodoulou A, Huang WE. Reprogramming synthetic cells for targeted cancer therapy. ACS Synth Biol 2022;11:1349–60. https://doi.org/10.1021/acssynbio.1c00631.

[123] Vincent RL, Gurbatri CR, Li F, Vardoshvili A, Coker C, Im J, et al. Probiotic-guided CAR-T cells for solid tumor targeting. Science 2023;382:211–18. https://doi.org/10.1126/science.add7034.

[124] Chapman MR, Robinson LS, Pinkner JS, Roth R, Heuser J, Hammar M, et al. Role of Escherichia coli curli operons in directing amyloid fiber formation. Science 2002;295:851–55. https://doi.org/10.1126/science.1067484.

[125] Barnhart MM, Chapman MR. Curli biogenesis and function. Annu Rev Microbiol 2006;60:131–47. https://doi.org/10.1146/annurev.micro.60.080805.142106.

[126] An B, Wang Y, Jiang X, Ma C, Mimee M, Moser F, et al. Programming living glue systems to perform autonomous mechanical repairs. Matter 2020;3:2080–92. https://doi.org/10.1016/j.matt.2020.09.006.

[127] Praveschotinunt P, Duraj-Thatte AM, Gelfat I, Bahl F, Chou DB, Joshi NS. Engineered E. coli Nissle 1917 for the delivery of matrix-tethered therapeutic domains to the gut. Nat Commun 2019;10. https://doi.org/10.1038/s41467-019-13336-6.

[128] Pu J, Liu Y, Zhang J, An B, Li Y, Wang X, et al. Virus disinfection from environmental water sources using living engineered biofilm materials. Adv Sci 2020;7:1903558. https://doi.org/10.1002/advs.201903558.

[129] Botyanszki Z, Tay PKR, Nguyen PQ, Nussbaumer MG, Joshi NS. Engineered catalytic biofilms: Site-specific enzyme immobilization onto E. coli curli nanofibers. Biotechnol Bioeng 2015;112:2016–24. https://doi.org/10.1002/bit.25638.

[130] Jiang L, Song X, Li Y, Xu Q, Pu J, Huang H, et al. Programming integrative extracellular and intracellular biocatalysis for rapid, robust, and recyclable synthesis of trehalose. ACS Catal 2018;8:1837–42. https://doi.org/10.1021/acscatal.7b03445.

[131] Kalyoncu E, Ahan RE, Olmez TT, Seker UOS. Genetically encoded conductive protein nanofibers secreted by engineered cells. RSC Adv 2017;7:32543–51. https://doi.org/10.1039/C7RA06289C.

[132] Böhning J, Ghrayeb M, Pedebos C, Abbas DK, Khalid S, Chai L, et al. Donor-strand exchange drives assembly of the TasA scaffold in Bacillus subtilis biofilms. Nat Commun 2022;13:7082. https://doi.org/10.1038/s41467-022-34700-z.

[133] Mammeri NE, Hierrezuelo J, Tolchard J, Cámara-Almirón J, Caro-Astorga J, Álvarez-Mena A, et al. Molecular architecture of bacterial amyloids in Bacillus biofilms. FASEB J 2019;33:12146–63. https://doi.org/10.1096/fj.201900831R.

[134] Romero D, Aguilar C, Losick R, Kolter R. Amyloid fibers provide structural integrity to Bacillus subtilis biofilms. Proc Natl Acad Sci 2010;107:2230–34. https://doi.org/10.1073/pnas.0910560107.

[135] Vogt CM, Schraner EM, Aguilar C, Eichwald C. Heterologous expression of antigenic peptides in Bacillus subtilis biofilms. Microb Cell Factories 2016;15:137. https://doi.org/10.1186/s12934-016-0532-5.

[136] Huang J, Liu S, Zhang C, Wang X, Pu J, Ba F, et al. Programmable and printable Bacillus subtilis biofilms as engineered living materials. Nat Chem Biol 2019;15:34–41. https://doi.org/10.1038/s41589-018-0169-2.

[137] Rodrigo-Navarro A, Rico P, Saadeddin A, Garcia AJ, Salmeron-Sanchez M. Living biointerfaces based on non-pathogenic bacteria to direct cell differentiation. Sci Rep 2014;4:5849. https://doi.org/10.1038/srep05849.

[138] Hay JJ, Rodrigo-Navarro A, Hassi K, Moulisova V, Dalby MJ, Salmeron-Sanchez M. Living biointerfaces based on non-pathogenic bacteria support stem cell differentiation. Sci Rep 2016;6:21809. https://doi.org/10.1038/srep21809.

[139] Saadeddin A, Rodrigo-Navarro A, Monedero V, Rico P, Moratal D, González-Martín ML, et al. Functional living biointerphases. Adv Healthc Mater 2013;2:1213–18. https://doi.org/10.1002/adhm.201200473.

[140] Li L, Yang C, Ma B, Lu S, Liu J, Pan Y, et al. Hydrogel-encapsulated engineered microbial consortium as a photoautotrophic "Living Material" for promoting skin wound healing. ACS Appl Mater Interfaces 2023;15:6536–47. https://doi.org/10.1021/acsami.2c20399.

[141] Tajima K, Imai T, Yui T, Yao M, Saxena I. Cellulose-synthesizing machinery in bacteria. Cellulose 2022;29:2755–77. https://doi.org/10.1007/s10570-021-04225-7.

[142] Jeong SI, Lee SE, Yang H, Jin Y-H, Park C-S, Park YS. Toxicologic evaluation of bacterial synthesized cellulose in endothelial cells and animals. Mol Cell Toxicol 2010;6:370–77. https://doi.org/10.1007/s13273-010-0049-7.

[143] Augimeri RV, Strap JL. The phytohormone ethylene enhances cellulose production, regulates CRP/FNRKx transcription and causes differential gene expression within the bacterial cellulose synthesis operon of Komagataeibacter (Gluconacetobacter) xylinus ATCC 53582. Front Microbiol 2015;6. https://doi.org/10.3389/fmicb.2015.01459.

[144] Schaffner M, Rühs PA, Coulter F, Kilcher S, Studart AR. 3D printing of bacteria into functional complex materials. Sci Adv 2017;3:eaao6804. https://doi.org/10.1126/sciadv.aao6804.

[145] Binelli MR, Rühs PA, Pisaturo G, Leu S, Trachsel E, Studart AR. Living materials made by 3D printing cellulose-producing bacteria in granular gels. Biomater Adv 2022;141:213095. https://doi.org/10.1016/j.bioadv.2022.213095.

[146] Caro-Astorga J, Walker KT, Herrera N, Lee K-Y, Ellis T. Bacterial cellulose spheroids as building blocks for 3D and patterned living materials and for regeneration. Nat Commun 2021;12:5027. https://doi.org/10.1038/s41467-021-25350-8.

[147] Gilbert C, Ellis T. Biological engineered living materials: Growing functional materials with genetically programmable properties. ACS Synth Biol 2019;8:1–15. https://doi.org/10.1021/acssynbio.8b00423.

[148] Gilbert C, Tang T-C, Ott W, Dorr BA, Shaw WM, Sun GL, et al. Living materials with programmable functionalities grown from engineered microbial co-cultures. Nat Mater 2021;20:691–700. https://doi.org/10.1038/s41563-020-00857-5.

[149] Rivera-Tarazona LK, Campbell ZT, Ware TH. Stimuli-responsive engineered living materials. Soft Matter 2021;17:785–809. https://doi.org/10.1039/D0SM01905D.

[150] Liu X, Tang T-C, Tham E, Yuk H, Lin S, Lu TK, et al. Stretchable living materials and devices with hydrogel–elastomer hybrids hosting programmed cells. Proc Natl Acad Sci 2017;114:2200–05. https://doi.org/10.1073/pnas.1618307114.

[151] Liu X, Yuk H, Lin S, Parada GA, Tang T-C, Tham E, et al. 3D printing of living responsive materials and devices. Adv Mater 2018;30:1704821. https://doi.org/10.1002/adma.201704821.

[152] Guo S, Dubuc E, Rave Y, Verhagen M, Twisk SAE, van der Hek T, et al. Engineered living materials based on adhesin-mediated trapping of programmable cells. ACS Synth Biol 2020;9:475–85. https://doi.org/10.1021/acssynbio.9b00404.

[153] Mimee M, Nadeau P, Hayward A, Carim S, Flanagan S, Jerger L, et al. An ingestible bacterial-electronic system to monitor gastrointestinal health. Science 2018;360:915–18. https://doi.org/10.1126/science.aas9315.

[154] Du H, Amstad E. Water: How does it influence the CaCO3 formation? Angew Chem Int Ed 2020;59:1798–816. https://doi.org/10.1002/anie.201903662.

[155] Wang Y, Von Euw S, Fernandes FM, Cassaignon S, Selmane M, Laurent G, et al. Water-mediated structuring of bone apatite. Nat Mater 2013;12:1144–53. https://doi.org/10.1038/nmat3787.

[156] Zhang J, Ji Y, Jiang S, Shi M, Cai W, Miron RJ, et al. Calcium–Collagen coupling is vital for biomineralization schedule. Adv Sci 2021;8:2100363. https://doi.org/10.1002/advs.202100363.

[157] Schulz A, Wang H, van Rijn P, Böker A. Synthetic inorganic materials by mimicking biomineralization processes using native and non-native protein functions. J Mater Chem 2011;21:18903–18. https://doi.org/10.1039/C1JM12490K.

[158] Kırpat Konak BM, Bakar ME, Ahan RE, Özyürek EU, Dökmeci S, Şafak Şeker UÖ. A living material platform for the biomineralization of biosilica. Mater Today Bio 2022;17:100461. https://doi.org/10.1016/j.mtbio.2022.100461.

[159] Duman E, Şahin Kehribar E, Ahan RE, Yuca E, Şeker UÖŞ. Biomineralization of calcium phosphate crystals controlled by protein–protein interactions. ACS Biomater Sci Eng 2019;5:4750–63. https://doi.org/10.1021/acsbiomaterials.9b00649.

[160] Wang Y, An B, Xue B, Pu J, Zhang X, Huang Y, et al. Living materials fabricated via gradient mineralization of light-inducible biofilms. Nat Chem Biol 2021;17:351–59. https://doi.org/10.1038/s41589-020-00697-z.

[161] Yan S, Zeng X, Wang Y, Liu B-F. Biomineralization of bacteria by a metal–organic framework for therapeutic delivery. Adv Healthc Mater 2020;9:2000046. https://doi.org/10.1002/adhm.202000046.

[162] Yavuz M, Ütkür M, Kehribar EŞ, Yağız E, Sarıtaş EÜ, Şeker UÖŞ. Engineered bacteria with genetic circuits accumulating nanomagnets as MRI contrast agents. Small 2022;18:2200537. https://doi.org/10.1002/smll.202200537.

[163] Keestra M, Rutting L, Post G, de Roo M, Blad S, de Greef L. An Introduction to Interdisciplinary Research: Theory and Practice, Amsterdam University Press; 2016.

[164] Kluger MO, Bartzke G. A practical guideline how to tackle interdisciplinarity – A synthesis from a post-graduate group project. Humanit Soc Sci Commun 2020;7:1–11. https://doi.org/10.1057/s41599-020-00540-9.

[165] Learning to think together: Creativity, interdisciplinary collaboration and epistemic control – ScienceDirect n.d. https://www.sciencedirect.com/science/article/abs/pii/S1871187120302236 (accessed June 14, 2024).

[166] Wang -C-C. Using design thinking for interdisciplinary curriculum design and teaching: a case study in higher education. Humanit Soc Sci Commun 2024;11:1–13. https://doi.org/10.1057/s41599-024-02813-z.

[167] Overview of DNA Microarrays: Types, Applications, and Their Future – Bumgarner – 2013 – Current Protocols in Molecular Biology – Wiley Online Library n.d. https://currentprotocols.onlinelibrary.wiley.com/doi/10.1002/0471142727.mb2201s101 (accessed June 14, 2024).

[168] Microarray technology: beyond transcript profiling and genotype analysis | Nature Reviews Genetics n.d. https://www.nature.com/articles/nrg1809 (accessed June 14, 2024).

[169] Shiu S-H, Borevitz JO. The next generation of microarray research: applications in evolutionary and ecological genomics. Heredity 2008;100:141–49. https://doi.org/10.1038/sj.hdy.6800916.

[170] Neelapu SS, Tummala S, Kebriaei P, Wierda W, Gutierrez C, Locke FL, et al. Chimeric antigen receptor T-cell therapy – assessment and management of toxicities. Nat Rev Clin Oncol 2018;15:47–62. https://doi.org/10.1038/nrclinonc.2017.148.

[171] Ducarmon QR, Kuijper EJ, Olle B. Opportunities and challenges in development of live biotherapeutic products to fight infections. J Infect Dis 2021;223:S283–9. https://doi.org/10.1093/infdis/jiaa779.

[172] Better by design: What to expect from novel CAR-engineered cell therapies? – ScienceDirect n.d. https://www.sciencedirect.com/science/article/abs/pii/S0734975022000131?via%3Dihub (accessed June 14, 2024).

[173] DeFrancesco L. The making of the Biohub. Nat Biotechnol 2020;38:1116–20. https://doi.org/10.1038/s41587-020-0685-y.

[174] The WHO BioHub system: experiences from the pilot phase | BMJ Global Health n.d. https://gh.bmj.com/content/8/8/e013421 (accessed June 14, 2024).

[175] The Chan-Zuckerberg Biohub: Modern Philanthrocapitalism Through a Critical Lens | The Columbia University Journal of Global Health n.d. https://journals.library.columbia.edu/index.php/jgh/article/view/6642 (accessed June 14, 2024).

[176] Check Hayden E. Facebook couple commits $3 billion to cure disease. Nature 2016;537:595–595. https://doi.org/10.1038/nature.2016.20649.

[177] UCSF establishes $250M initiative to develop living therapeutics – The Cancer Letter n.d. https://cancerletter.com/in-brief/20210625_8h/ (accessed June 14, 2024).

[178] Living Therapeutics Initiative Will Accelerate Development and Delivery of Revolutionary Treatments 2021.

[179] Living Therapeutics Initiative U. Thermo Fisher partners with UCSF to provide a one-stop shop for early biotech manufacturing. News Release 2021. https://livingcelltherapy.ucsf.edu/news/thermo-fisher-partners-ucsf-provide-one-stop-shop-early-biotech-manufacturing (accessed May 30, 2024).

[180] Report S. Stanford launches major effort to harness the microbiome to treat disease. News Release 2019. https://news.stanford.edu/stories/2019/08/stanford-launches-major-effort-harness-microbiome-treat-disease (accessed May 30, 2024).

[181] Michael Fischbach wants to build a microbiome from the ground up. So what do we need to get there? News 2019. https://chemh.stanford.edu/news/michael-fischbach-wants-build-microbiome-ground-so-what-do-we-need-get-there (accessed May 30, 2024).

[182] Novo Nordisk. New cell therapy facility enters fight against chronic diseases. News Release 2023. https://novonordiskfonden.dk/en/news/new-cell-therapy-facility-enters-fight-against-chronic-diseases/ (accessed May 30, 2024).

[183] Cowley AB. Navigating microbiome innovations: Strategic partnerships in Live Biotherapeutic product development 2024. https://www.microbiometimes.com/navigating-microbiome-innovations-strategic-partnerships-in-live-biotherapeutic/ (accessed May 27, 2024).

[184] Khraishi M, Stead D, Lukas M, Scotte F, Schmid H. Biosimilars: A multidisciplinary perspective. Clin Ther 2016;38:1238–49. https://doi.org/10.1016/j.clinthera.2016.02.023.

[185] From CMO to CDMO: Opportunities for Specializing and Innovation n.d. https://www.bioprocessintl.com/contract-services/from-cmo-to-cdmo-opportunities-for-specializing-and-innovation (accessed June 14, 2024).

[186] Contract development and manufacturing organization selection: critical considerations that can make or break your cell and gene therapy development – Cytotherapy n.d. https://www.isct-cytotherapy.org/article/S1465-3249(24)00090-2/abstract (accessed June 14, 2024).

[187] Sunday, January 31, Tweet 2016. Putting the "D" in CDMO with Advanced Process Development n.d. https://www.americanpharmaceuticalreview.com/Featured-Articles/182916-Putting-the-D-in-CDMO-with-Advanced-Process-Development/ (accessed June 14, 2024).

[188] Spacova I, Binda S, Ter Haar JA, Henoud S, Legrain-Raspaud S, Dekker J, et al. Comparing technology and regulatory landscape of probiotics as food, dietary supplements and live biotherapeutics. Front Microbiol 2023;14:1272754. https://doi.org/10.3389/fmicb.2023.1272754.

[189] Bellino S, La Salvia A, Cometa MF, Botta R. Cell-based medicinal products approved in the European Union: current evidence and perspectives. Front Pharmacol 2023;14:1200808. https://doi.org/10.3389/fphar.2023.1200808.

[190] Microbial Ecology in Health and Disease. Taylor Francis n.d. https://www.tandfonline.com/journals/zmeh20 (accessed June 14, 2024).

[191] Franciosa G, Guida S, Gomez Miguel MJ, Von Hunolstein C. Live biotherapeutic products and their regulatory framework in Italy and Europe. Ann Ist Super Sanita 2023;59:56–67. https://doi.org/10.4415/ANN_23_01_09.

[192] Rouanet A, Bolca S, Bru A, Claes I, Cvejic H, Girgis H, et al. Live biotherapeutic products, a road map for safety assessment. Front Med 2020;7. https://doi.org/10.3389/fmed.2020.00237.

[193] Cordaillat-Simmons M, Rouanet A, Pot B. Live biotherapeutic products: the importance of a defined regulatory framework. Exp Mol Med 2020;52:1397–406. https://doi.org/10.1038/s12276-020-0437-6.

[194] Dreher-Lesnick SM, Stibitz S, Carlson PE Jr. U.S. regulatory considerations for development of live biotherapeutic products as drugs. Microbiol Spectr 2017;5:5.5.11. https://doi.org/10.1128/microbiol spec.BAD-0017-2017.

[195] Food and Drug Administration. Early Clinical Trials with Live Biotherapeutic Products: Chemistry, Manufacturing, and Control Information. Guid Ind 2016. https://www.fda.gov/files/vaccines,% 20blood%20%26%20biologics/published/Early-Clinical-Trials-With-Live-Biotherapeutic-Products–Chemistry–Manufacturing–and-Control-Information–Guidance-for-Industry.pdf (accessed May 31, 2024).

[196] Cordaillat-Simmons M, Rouanet A, Pot B. Food or Pharma: The Name does Make a Difference. Probiotics Prev. Manag. Hum. Dis., Elsevier; 2022, pp. 13–40. https://doi.org/10.1016/B978-0-12-823733-5.00005-2.

[197] Katkowska M, Garbacz K, Kusiak A. Probiotics: Should all patients take them? Microorganisms 2021;9:2620. https://doi.org/10.3390/microorganisms9122620.

[198] Ferring, Rebiotix and MyBiotics Collaborate to Develop Live Microbiome-Based Therapeutics in Reproductive Medicine and Maternal Health – Ferring Global n.d. https://www.ferring.com/ferring-rebiotix-and-mybiotics-collaborate-to-develop-live-microbiome-based-therapeutics-in-reproductive-medicine-and-maternal-health/ (accessed June 14, 2024).

[199] immunocity. Microbiotica Announces Clinical Trial Collaboration with MSD to Evaluate MB097 in Combination with KEYTRUDA® (pembrolizumab) in a Phase 1b Clinical Trial in Melanoma – Microbiotica n.d. https://microbiotica.com/microbiotica-announces-clinical-trial-collaboration-with-msd-to-evaluate-mb097-in-combination-with-keytruda-pembrolizumab-in-a-phase-1b-clinical-trial-in-melanoma/ (accessed June 14, 2024).

[200] Robinson MJ, Vervier K, Harris S, Popple A, Klisko D, Hudson R, et al. Abstract P074: MB097: A therapeutic consortium of bacteria clinically-defined by precision microbiome profiling of immune checkpoint inhibitor patients with potent anti-tumor efficacy in vitro and in vivo. Cancer Immunol Res 2022;10:P074–P074. https://doi.org/10.1158/2326-6074.TUMIMM21-P074.

[201] Inc. L. Bacthera and Seres Therapeutics Collaborate for Commercial Manufacturing of SER-109, a Potential Treatment Against Recurrent C. difficile Infection 2024. https://www.lonza.com/news/2021-11-10-08-00 (accessed May 27, 2024).

[202] Feuerstadt P, Louie TJ, Lashner B, Wang EEL, Diao L, Bryant JA, et al. SER-109, an oral microbiome therapy for recurrent *Clostridioides* difficile infection. N Engl J Med 2022;386:220–29. https://doi.org/10.1056/NEJMoa2106516.

[203] McGovern BH, Ford CB, Henn MR, Pardi DS, Khanna S, Hohmann EL, et al. SER-109, an investigational microbiome drug to reduce recurrence after *Clostridioides difficile* infection: Lessons learned from a phase 2 trial. Clin Infect Dis 2021;72:2132–40. https://doi.org/10.1093/cid/ciaa387.

[204] Lilly E Lilly and Sigilon Therapeutics Announce Strategic Collaboration to Develop Encapsulated Cell Therapies for the Treatment of Type 1 Diabetes 2018. https://investor.lilly.com/news-releases/news-release-details/lilly-and-sigilon-therapeutics-announce-strategic-collaboration (accessed May 27, 2024).

[205] First cell therapy for diabetes approved. Nat Biotechnol 2023;41:1036–1036. https://doi.org/10.1038/s41587-023-01912-7.

[206] Vegas AJ, Veiseh O, Doloff JC, Ma M, Tam HH, Bratlie K, et al. Combinatorial hydrogel library enables identification of materials that mitigate the foreign body response in primates. Nat Biotechnol 2016;34:345–52. https://doi.org/10.1038/nbt.3462.

[207] Kelly O, McClatchey PM, Bawadekar M, Higgins J, Barrett K, Buchanan M, et al. 401.5: Shielded living therapeutics platform for the treatment of type 1 diabetes. Transplantation 2023;107:160–160. https://doi.org/10.1097/01.tp.0000994648.23711.e3.

[208] Lilly E Lilly to Acquire Sigilon Therapeutics 2023. https://investor.lilly.com/news-releases/news-release-details/lilly-acquire-sigilon-therapeutics (accessed May 27, 2024).

[209] Allegretti JR CP101 for the Treatment of Ulcerative Colitis. clinicaltrials.gov 2024.

[210] Inc. F. Finch Therapeutics Announces Clinical Collaboration in Ulcerative Colitis with Brigham and Women's Hospital and Updates to University of Minnesota License Agreement. Press Release 2024. https://ir.finchtherapeutics.com/news-releases/news-release-details/finch-therapeutics-announces-clinical-collaboration-ulcerative.

[211] Van Der Lelie D, Oka A, Taghavi S, Umeno J, Fan T-J, Merrell KE, et al. Rationally designed bacterial consortia to treat chronic immune-mediated colitis and restore intestinal homeostasis. Nat Commun 2021;12:3105. https://doi.org/10.1038/s41467-021-23460-x.

[212] Qiu P, Ishimoto T, Fu L, Zhang J, Zhang Z, Liu Y. The gut microbiota in inflammatory bowel disease. Front Cell Infect Microbiol 2022;12. https://doi.org/10.3389/fcimb.2022.733992.

[213] Global G Nature Communications – Rationally Design Bacterial Consortia To Treat Chronic Immune Medicated Colitis And Restore Intestinal Homeoestasis n.d. https://gustoglobal.com/nature-communications-rationally-design-bacterial-consortia-to-treat-chronic-immune-medicated-colitis-and-restore-intestinal-homeoestasis/.

[214] Inc. S. Synlogic Enters Research Collaboration with Roche for Development of Novel Therapy to Treat Inflammatory Bowel Disease. Press Release 2021. https://investor.synlogictx.com/news-releases/news-release-details/synlogic-enters-research-collaboration-roche-development-novel (accessed May 28, 2024).

[215] Spark Therapeutics Enters Collaboration with Senti Bio to Bolster Industry-Leading Gene Therapy Research Platform – Spark Therapeutics n.d. https://sparktx.com/press_releases/senti-collaboration/ (accessed June 14, 2024).

[216] Nissim L, Wu M-R, Pery E, Binder-Nissim A, Suzuki HI, Stupp D, et al. Synthetic RNA-based immunomodulatory gene circuits for cancer immunotherapy. Cell 2017;171:1138–50, e15. https://doi.org/10.1016/j.cell.2017.09.049.

[217] Schukur L, Geering B, Charpin-El Hamri G, Fussenegger M. Implantable synthetic cytokine converter cells with AND-gate logic treat experimental psoriasis. Sci Transl Med 2015;7. https://doi.org/10.1126/scitranslmed.aac4964.

[218] Bio S Senti Bio Announces New Strategic Collaboration with Celest Therapeutics for Clinical Development of SENTI-301A in China. Press Release 2023. https://sentibio.gcs-web.com/news-releases/news-release-details/senti-bio-announces-new-strategic-collaboration-celest (accessed May 29, 2024).

[219] Bioworks G Microba Life Sciences and Ginkgo Bioworks Announce Partnership to Discover Novel Live Biotherapeutics. Press Release 2022. https://investors.ginkgobioworks.com/news/news-details/2022/Microba-Life-Sciences-and-Ginkgo-Bioworks-Announce-Partnership-to-Discover-Novel-Live-Biotherapeutics/default.aspx (accessed May 30, 2024).

[220] Pribyl AL, Parks DH, Angel NZ, Boyd JA, Hasson AG, Fang L, et al. Critical evaluation of faecal microbiome preservation using metagenomic analysis. ISME Commun 2021;1:14. https://doi.org/10.1038/s43705-021-00014-2.

[221] Parks DH, Rigato F, Vera-Wolf P, Krause L, Hugenholtz P, Tyson GW, et al. Evaluation of the microba community profiler for taxonomic profiling of metagenomic datasets from the human gut microbiome. Front Microbiol 2021;12. https://doi.org/10.3389/fmicb.2021.643682.

[222] AbbVie and Umoja Biopharma Announce Strategic Collaboration to Develop Novel In-Situ CAR-T Cell Therapies. Press Release 2024. https://news.abbvie.com/2024-01-04-AbbVie-and-Umoja-Biopharma-Announce-Strategic-Collaboration-to-Develop-Novel-In-Situ-CAR-T-Cell-Therapies (accessed May 30, 2024).

[223] Michels KR, Sheih A, Hernandez SA, Brandes AH, Parrilla D, Irwin B, et al. Preclinical proof of concept for VivoVec, a lentiviral-based platform for in vivo CAR T-cell engineering. J Immunother Cancer 2023;11:e006292. https://doi.org/10.1136/jitc-2022-006292.

[224] Therapeutics B BlueRock Therapeutics and bit.bio announce collaboration and option agreement for the discovery and manufacture of regulatory T cell (Treg) based therapies. Press Release 2023. https://www.bluerocktx.com/bluerock-therapeutics-and-bit-bio-announce-collaboration-and-option-agreement-for-the-discovery-and-manufacture-of-regulatory-t-cell-treg-based-therapies/ (accessed May 30, 2024).

[225] Pawlowski M, Ortmann D, Bertero A, Tavares JM, Pedersen RA, Vallier L, et al. Inducible and deterministic forward programming of human pluripotent stem cells into neurons, skeletal myocytes, and oligodendrocytes. Stem Cell Rep 2017;8:803–12. https://doi.org/10.1016/j.stemcr.2017.02.016.

[226] Pavlinek A, Matuleviciute R, Sichlinger L, Dutan Polit L, Armeniakos N, Vernon AC, et al. Interferon-γ exposure of human iPSC-derived neurons alters major histocompatibility complex I and synapsin protein expression. Front Psychiatry 2022;13. https://doi.org/10.3389/fpsyt.2022.836217.

[227] McCune JM, Stevenson SC, Doehle BP, Trenor CC, Turner EH, Spector JM. Collaborative science to advance gene therapies in resource-limited parts of the world. Mol Ther 2021;29:3101–02. https://doi.org/10.1016/j.ymthe.2021.05.024.

[228] NIH. Ending the HIV Epidemic in the U.S. (EHE). Press Release 2023. https://www.hiv.gov/federal-response/ending-the-hiv-epidemic/overview (accessed May 28, 2024).

[229] NIH launches new collaboration to develop gene-based cures for sickle cell disease and HIV on global scale. News Release 2019. https://www.nih.gov/news-events/news-releases/nih-launches-new-collaboration-develop-gene-based-cures-sickle-cell-disease-hiv-global-scale (accessed May 29, 2024).

[230] Artificial intelligence and synthetic biology: A tri-temporal contribution – ScienceDirect n.d. https://www.sciencedirect.com/science/article/abs/pii/S0303264716300016?via%3Dihub (accessed June 14, 2024).

[231] Parallel networks: Synthetic biology and artificial intelligence. ACM J Emerg Technol Comput Syst n.d.;11(3). https://dl.acm.org/doi/10.1145/2667229 (accessed June 14, 2024).

[232] Artificial intelligence for synthetic biology | Communications of the ACM n.d. https://dl.acm.org/doi/10.1145/3500922 (accessed June 14, 2024).

[233] Häse F, Roch LM, Aspuru-Guzik A. Next-generation experimentation with self-driving laboratories. Trends Chem 2019;1:282–91. https://doi.org/10.1016/j.trechm.2019.02.007.

[234] Driving school for self-driving labs – Digital Discovery (RSC Publishing) n.d. https://pubs.rsc.org/en/content/articlelanding/2023/dd/d3dd00150d (accessed June 14, 2024).

[235] Abolhasani M, Kumacheva E. The rise of self-driving labs in chemical and materials sciences. Nat Synth 2023;2:483–92. https://doi.org/10.1038/s44160-022-00231-0.

[236] Biosystems Design by Machine Learning | ACS Synthetic Biology n.d. https://pubs.acs.org/doi/10.1021/acssynbio.0c00129 (accessed June 14, 2024).

[237] Choi KR, Jang WD, Yang D, Cho JS, Park D, Lee SY. Systems metabolic engineering strategies: Integrating systems and synthetic biology with metabolic engineering. Trends Biotechnol 2019;37:817–37. https://doi.org/10.1016/j.tibtech.2019.01.003.

[238] Mohmad Yousoff SN, Baharin A, Abdullah A. Differential Search Algorithm in Deep Neural Network for the Predictive Analysis of Xylitol Production in Escherichia Coli. In: Mohamed Ali MS, Wahid H, Mohd Subha NA, Sahlan S, Md. Yunus MA, Wahap AR, editors. Model. Des. Simul. Syst., Singapore: Springer; 2017, pp. 53–67. https://doi.org/10.1007/978-981-10-6502-6_5.

[239] Systems biology approaches integrated with artificial intelligence for optimized metabolic engineering – ScienceDirect n.d. https://www.sciencedirect.com/science/article/pii/S2214030120300493?via%3Dihub (accessed June 14, 2024).

[240] Zhang J, Petersen SD, Radivojevic T, Ramirez A, Pérez-Manríquez A, Abeliuk E, et al. Combining mechanistic and machine learning models for predictive engineering and optimization of tryptophan metabolism. Nat Commun 2020;11:4880. https://doi.org/10.1038/s41467-020-17910-1.

[241] Salehi M, Farhadi S, Moieni A, Safaie N, Ahmadi H. Mathematical modeling of growth and paclitaxel biosynthesis in Corylus avellana cell culture responding to fungal elicitors using multilayer perceptron-genetic algorithm. Front Plant Sci 2020;11. https://doi.org/10.3389/fpls.2020.01148.

[242] Guiding the Refinement of Biochemical Knowledgebases with Ensembles of Metabolic Networks and Machine Learning: Cell Systems n.d. https://www.cell.com/cell-systems/fulltext/S2405-4712(19) 30392-8?_returnURL=https%3A%2F%2Flinkinghub.elsevier.com%2Fretrieve%2Fpii% 2FS2405471219303928%3Fshowall%3Dtrue (accessed June 14, 2024).

[243] New synthetic biology tools for metabolic control – ScienceDirect n.d. https://www.sciencedirect. com/science/article/pii/S0958166922000581?via%3Dihub (accessed June 14, 2024).

[244] Skylar-Scott MA, Uzel SGM, Nam LL, Ahrens JH, Truby RL, Damaraju S, et al. Biomanufacturing of organ-specific tissues with high cellular density and embedded vascular channels. Sci Adv 2019;5: eaaw2459. https://doi.org/10.1126/sciadv.aaw2459.

[245] Unterholzner SJ, Poppenberger B, Rozhon W. Toxin-antitoxin systems: Biology, identification, and application. Mob Genet Elem 2013;3:e26219. https://doi.org/10.4161/mge.26219.

[246] Nyerges A, Vinke S, Flynn R, Owen SV, Rand EA, Budnik B, et al. A swapped genetic code prevents viral infections and gene transfer. Nature 2023;615:720–27. https://doi.org/10.1038/s41586-023-05824-z.

[247] Zürcher JF, Robertson WE, Kappes T, Petris G, Elliott TS, Salmond GPC, et al. Refactored genetic codes enable bidirectional genetic isolation. Science 2022;378:516–23. https://doi.org/10.1126/sci ence.add8943.

[248] U.S. FDA O of the. Early Clinical Trials With Live Biotherapeutic Products: Chemistry, Manufacturing, and Control Information 2021.

[249] Guallar-Garrido S, Julián E. Bacillus Calmette-Guérin (BCG) therapy for bladder cancer: An update. ImmunoTargets Ther 2020;9:1–11. https://doi.org/10.2147/ITT.S202006.

[250] Zamat A, Zhu L, Wang Y. Engineering molecular machines for the control of cellular functions for diagnostics and therapeutics. Adv Funct Mater 2020;30:1904345. https://doi.org/10.1002/adfm. 201904345.

[251] Lipsitz YY, Bedford P, Davies AH, Timmins NE, Zandstra PW. Achieving efficient manufacturing and quality assurance through synthetic cell therapy design. Cell Stem Cell 2017;20:13–17. https://doi. org/10.1016/j.stem.2016.12.003.

[252] Lipsitz YY, Timmins NE, Zandstra PW. Quality cell therapy manufacturing by design. Nat Biotechnol 2016;34:393–400. https://doi.org/10.1038/nbt.3525.

[253] Lentini R, Santero SP, Chizzolini F, Cecchi D, Fontana J, Marchioretto M, et al. Integrating artificial with natural cells to translate chemical messages that direct E. coli behaviour. Nat Commun 2014;5:4012. https://doi.org/10.1038/ncomms5012.

[254] Gardner TJ, Bourne CM, Dacek MM, Kurtz K, Malviya M, Peraro L, et al. Targeted cellular micropharmacies: Cells engineered for localized drug delivery. Cancers 2020;12:2175. https://doi. org/10.3390/cancers12082175.

[255] Li Z, Hu S, Cheng K. Chemical engineering of cell therapy for heart diseases. Acc Chem Res 2019;52:1687–96. https://doi.org/10.1021/acs.accounts.9b00137.

[256] Kim TH, Ju K, Kim SK, Woo S-G, Lee J-S, Lee C-H, et al. Novel signal peptides and episomal plasmid system for enhanced protein secretion in engineered bacteroides species. ACS Synth Biol 2024;13:648–57. https://doi.org/10.1021/acssynbio.3c00649.

[257] Prox J, Smith T, Holl C, Chehade N, Guo L. Integrated biocircuits: engineering functional multicellular circuits and devices. J Neural Eng 2018;15:023001. https://doi.org/10.1088/1741-2552/ aaa906.

[258] Carlo DD. Technologies for the directed evolution of cell therapies. SLAS Technol 2019;24:359–72. https://doi.org/10.1177/2472630319834897.

[259] Riglar DT, Silver PA. Engineering bacteria for diagnostic and therapeutic applications. Nat Rev Microbiol 2018;16:214–25. https://doi.org/10.1038/nrmicro.2017.172.

[260] Huang Y, Lin X, Yu S, Chen R, Chen W. Intestinal engineered probiotics as living therapeutics: Chassis selection, colonization enhancement, gene circuit design, and biocontainment. ACS Synth Biol 2022;11:3134–53. https://doi.org/10.1021/acssynbio.2c00314.

[261] Chowdhury S, Castro S, Coker C, Hinchliffe TE, Arpaia N, Danino T. Programmable bacteria induce durable tumor regression and systemic antitumor immunity. Nat Med 2019;25:1057–63. https://doi.org/10.1038/s41591-019-0498-z.

[262] Li Y, Wang M, Hong S. Live-cell glycocalyx engineering. ChemBioChem 2023;24:e202200707. https://doi.org/10.1002/cbic.202200707.

[263] Yang S, Choi H, Nguyen DT, Kim N, Rhee SY, Han SY, et al. Bioempowerment of therapeutic living cells by single-cell surface engineering. Adv Ther 2023;6:2300037. https://doi.org/10.1002/adtp.202300037.

[264] Sarbadhikary P, George BP, Abrahamse H. Paradigm shift in future biophotonics for imaging and therapy: Miniature living lasers to cellular scale optoelectronics. Theranostics 2022;12:7335–50. https://doi.org/10.7150/thno.75905.

Index

3D printing technology 138

AND gate 19–21
anticancer agents 128
atopit dermatitis 123, 153

bacterial cellulose 138
bacterial strains 87, 95–96
biofabricating 151
biofilm 133–137, 142
Biohub 145
bioinformatics 144, 149
biological DNA transfer methods 7
biomarkers 107–108
biomineralization 141, 143
biopharmaceutical manufacturing 61
biophotonics 152

cancer detection methods 104
cancer therapy 87, 92–93, 95–97
CAR-T 73, 75, 77
cell line development 62
cell receptors 120
cell-fate decisions 40
cellular decision-making 38, 41–42
cellular decisions 41
cellular engineering 150
chimeric antigen receptor 33
circuit-engineered bacteria 130
Coley toxins 128
coordinated overexpression 64
CRISPR-based approaches 127
curli operon 133–135
cytokines 120

de novo DNA synthesis strategies 4
decision-making of cellular processes 38
DNA encoding 3–4
DNA sequencing 88

E. coli 91, 94, 98–102, 106
E. coli Nissle 124
electrical stimulation 51
electron transport chain 74
EMeRALD 11
endoplasmic reticulum 60, 77

engineer workhorse microorganisms 129
engineered bacteria 130
engineered cells 12
engineered living materials 132–133, 139
engineering cells 2, 9–10
enzymes 120, 134, 139
exosomes 64, 71–72
extracellular vesicles 71, 78

fecal microbiota transplantation 92, 121

gastrointestinal tract 88, 94, 104
GenScript 5
glycosylation 65–67, 69, 77
good manufacturing practices 43
gram-negative bacteria 101
gut microbiome 121–122, 125, 127–128, 131, 145, 147–148, 150

H. pylori 90
heat-activated gene regulation systems 51
heme-binding proteins 36

iDA reaction 4, 6
immune checkpoint therapy 125
inflammatory bowel disease 120, 122–123, 134, 148, 153
Integrated DNA Technologies 5
intracellular vesicle trafficking 64
intratumoral bacteria 129
intratumoral bioreactors 93
isothermal DNA assembly 4

knockdown 64, 68, 77–78
knockout 64, 68, 77–78

light-induced regulation of gene transcription 47
live biotherapeutic products 145–146
living therapeutics 119–121, 124–126, 128, 130–131, 145–151, 153

memory circuits 13
mesenchymal stem cells 72
microbe-based cancer therapeutics 107–108
microbiome 87–95, 106
microbiota-derived compounds 92

molecular cloning 3, 5

N-glycosylation 66, 68–69, 77
NOT gate 19

O-glycosylation 67, 69–70, 77
optogenetic tools 36
OR gate 19–20

phenylketonuria 127
photocaging system 49
photoswitch systems 48
physical DNA transfer methods 6
plasmid vector 5
preclinical testing 106
probiotic bacterium 36–37
prokaryotic and eukaryotic systems 41
protein-based contrast agents 36

recombinases 21

Salmonella 91, 94, 96, 98, 101–102
signal peptides 61–62, 78
signal processing 1, 3, 13–14, 17, 20, 22
Streptococcus 89
SYNB1891 105
synthetic biology 3, 5–6, 10–15, 17, 19–22, 119, 121,
         126–127, 134, 139, 142, 152–153
synthetic biology tools 31–34, 42, 52
synthetic genetic modules 8

thermally inducible promoters 102
transcription control 44
transcriptional sensing 9
transformative technologies 146
tumor invasion 101
tumor-specific internal signals 98
Twist Biosciences 5

urobiome 123
US Food and Drug Administration 126

www.ingramcontent.com/pod-product-compliance
Lightning Source LLC
Chambersburg PA
CBHW081529220326
41598CB00036B/6380